Urban Communication Regulation

URBAN communication

Gary Gumpert, *General Editor*

Volume 5

The Urban Communication series is part of the Peter Lang
Media and Communication list.
Every volume is peer reviewed and meets
the highest quality standards for content and production.

PETER LANG
New York • Bern • Frankfurt • Berlin
Brussels • Vienna • Oxford • Warsaw

Urban Communication Regulation

Communication Freedoms and Limits

Edited by
Harvey Jassem and Susan J. Drucker

PETER LANG
New York • Bern • Frankfurt • Berlin
Brussels • Vienna • Oxford • Warsaw

Library of Congress Cataloging-in-Publication Data

Names: Jassem, Harvey, editor. | Drucker, Susan J., editor.
Title: Urban communication regulation: communication freedoms and limits /
edited by Harvey Jassem and Susan J. Drucker.
Description: New York: Peter Lang, 2018.
Series: Urban communication; vol. 5 | ISSN 2153-1404
Includes bibliographical references and index.
Identifiers: LCCN 2017035224 | ISBN 978-1-4331-4631-2 (pbk.: alk. paper) |
ISBN 978-1-4331-4632-9 (hardback: alk. paper) |
ISBN 978-1-4331-4633-6 (ebook pdf) | ISBN 978-1-4331-4634-3 (epub) |
ISBN 978-1-4331-4635-0 (mobi)
Subjects: LCSH: Mass media policy. | Communication policy. |
Municipal government.
Classification: LCC P95.8.U73 | DDC 302.23—dc23
LC record available at https://lccn.loc.gov/2017035224
DOI 10.3726/b11474

Bibliographic information published by **Die Deutsche Nationalbibliothek**.
Die Deutsche Nationalbibliothek lists this publication in the "Deutsche
Nationalbibliografie"; detailed bibliographic data are available
on the Internet at http://dnb.d-nb.de/.

29 Broadway, 18th floor, New York, NY 10006
www.peterlang.com

Table of Contents

Figures

Introduction

HARVEY JASSEM
Professor Emeritus, University of Hartford

SUSAN J. DRUCKER
Hofstra University

Urban communication regulation is about as compelling and important an area for examination as we can think of. The 21st century has been dubbed the urban century. It is an era marked by recentralization and a return to the inner core after decades of suburban sprawl. It is an era of urban sprawl. It is the first century in which the majority of the world's people will live in urban areas with over three billion residents in cities representing a demographic transformation on an unprecedented scale. Cities have always been social settings.

As this is being written, people around the world are proclaiming that these are times of profound change and instability. Of course, such has often been said, and is often accurate. There are issues of political, economic, structural, climate, military, patriotic, educational, and racial concerns that are on the minds and in the voices of many. And how some of those and other concerns play out gets us back to urban communication regulation. At least if one believes, as we do, that decision making and the life of the mind depends on, and is reflective of, what people are aware of, thinking about, and acting on or choosing.

Nicholas Johnson, a former Commissioner of the Federal Communications Commission, noted that "The choice you'll never know is the choice you'll never make." (Johnson, 1972, p. 45) Knowing entails communication. So, to know, to understand, and to act, requires communication with other human beings. Increasingly, human beings are urban. By definition, "urban" is where the masses of people are.

Most people in the United States, the focus of this volume, and most people in the world, live in urban areas. That concentration is forecast to grow in the coming decades. So, if we want to know what most people are thinking about, if we want to know what most people can think about, a focus on urban is essential.

More than that, urban areas, because of their concentration of people and the constant mix of people, present special communication-related issues. Consider the communication issues that are purely density-related. If one lives in a very rural area, miles from another human being, would it matter if they crank the volume of their music such that it could be heard 500 feet away? Not at all. Would it matter if they put a large sign or flag or graffiti or a food stand outside their home? Hardly. Assuming there were a sidewalk outside their home, would they impede traffic if they stood there in a costume or playing an instrument asking passers by for money? We don't think so. But in a densely populated area, an urban area, all of these same things could be troublesome. And when people are bothered, they seek help. Government regulation is one avenue for such help.

Similarly, communication in urban areas involves infrastructure, more so than is likely the case in rural areas. Infrastructures and designated public spaces are more critical to life in an urban area than is the case in rural. Rural areas don't have the same need as urban ones for designated public outdoor spaces. There's already a lot of space. Rural areas tend not to need zoning or design standards because one person's design will hardly be seen or impact others. Funds for public art are likely slim, as funds tend to come from taxing individuals and businesses, and if there aren't many people, and they are spread out over a large area, financial resources likely do not run deep. And there are even differences in the physical infrastructures. One reason telephony became a regulated "natural monopoly" was that in the early days of telephony, when there were many competing telephone companies, the enormous amount of wiring that was necessary to handle all of those customers and competing companies in cities blackened the urban sky or clogged the underground conduits. Rural telephone poles didn't have that issue. And there are no rural underground conduits.

Urban areas differ from rural too due to their being the hub of multiculturalism. Immigrants bring different languages, foods, customs, and cultures to cities, adding to and sometimes clashing with, or at least not fully aware of some of the preexisting norms. There is the cacophony of rich and poor, young and old, residents and tourists, business districts and residential, vehicular and pedestrian traffic, crammed closely together. While all of that is exciting and can be fun, there are just more opportunities for misunderstandings, more opportunities for people to get in each other's ways, and for conflict.

And regulation, at its core, is about resource allocation and conflict resolution. Regulation is about setting up the rules of the game. Regulation is the result of choices and, affects the choices others will make. Regulation creates winners and losers—not just in terms of individuals or businesses, but in terms of arts, culture, ideas, politics and everything.

Regulation may be restrictive or expansive, negative or positive. By way of regulation, and we use the concept broadly in this volume, we refer to the roles governments play in affecting the communication landscape of cities. Sometime governments regulate who can say what to whom through what channels and under what conditions. In such cases, governments empower or disempower some "speakers." Governments may limit what may be "said" and those limits might be applied broadly or applied narrowly. Governments are also in positions where they may assign or limit the channels of communication. But it is not just about government limiting communication. Government policies or regulations can also serve to encourage or pave the way for some kinds of communication. Government grants for the arts or for infrastructure or for technological innovation can all serve to impact who says what to whom.

Some regulations directly control free expression in cities through licensing and control of content. Perhaps more interesting are the regulations which indirectly control urban communication, ostensibly targeting non-communication related activities but that purposefully or accidentally determine the nature of communicative interaction. Regulations are of two types: a) primary regulation: regulation primarily aimed at communicative activity and b) secondary regulation: regulations aimed at non-communicative activity but which affect communication. These secondary regulations include anti-loitering, anti-littering, gambling and public nuisances and the licensing of alcoholic beverages, minimum drinking age and driving age, parking regulations, anti-babble locations prohibiting the use of mobile phones. Smoking laws are among the most significant indirect regulators of communication opportunities in public places. Land use regulations, especially zoning laws are important secondary regulations (e.g. addressing signage, facades, display windows, take-out and delivery service, and vending machines, and the segregation of functions or design districts) all influence communication patterns. Additionally, much of the communication infrastructure is shaped by diverse regulations.

Communication regulation and regulatory theory also must deal with notions of jurisdiction and property rights. Who owns a place and does it matter? Do private property owners have rights to communicate in ways that travel beyond their property? Do they have the right to curtail communication when their spaces are open to the public? Do they have the right not to communicate?

"Regulations and licensing governing activities within the built environment influence communication by controlling communication activities or by controlling communication contexts" (Drucker & Gumpert, 1996). In 1996 Drucker and Gumpert argued there are "manifest and latent ways in which regulations governing both public spaces and the behavior in those spaces function as a unique, often overlooked area of communication law" (Drucker & Gumpert, 1996, p. 281). At that time they noted:

> Laurence Tribe (1978) has suggested that the regulation of communication is effectuated primarily through regulation aimed at communicative activities, or at communicative impact. Civil and criminal statutory law, common law, administrative rules and regulations, and executive orders are all sources of communication law. (Drucker & Gumpert, p. 281)

A complex regulatory web constitutes the regulation of communicative activities and impact.

There is a growing literature on urban communication and media and cosmopolitanism Featuring cosmopolitanism is featured by such authors as Myria Georgious's *Media and the City: Cosmopolitanism and Difference* (Georgiou, 2013) and Alexa Robertson's *A Mediated Cosmopolitanism: The World of Television News* (Robertson, 2010).

There has long been a literature on municipal law which addresses communication among myriad other areas of regulatory concern. More recently, several books have addressed communication rights in cities such as *Speech Out of Doors: Preserving First Amendment Liberties in Public Places* by Timothy Zick which deals with changes in the legal status of lands once available for the exercise of communicative rights in outdoor public spaces. Other volumes have dealt with the implications of privatization of public space. *Brave New Neighborhoods: The Privatization of Public*, by Margaret Kohn (2004) deals with First Amendment rights in public spaces which are becoming increasingly privatized and *Beyond Zuccotti Park: Freedom of Assembly and the Occupation of Public Space* by Ronald Shiffman et al. (2012) which deals with the legal status of public spaces in different countries as sites for public protest. A growing literature is concerned with smart cities (e.g.: *Smart Cities: Big Data, Civic Hackers, and the Quest for a New Utopia* by Anthony Townsend (2013). Several volumes address the growth of the wired or smart city which includes provision of municipal wi-fi. While policy and regulations (often retarding development) are addressed, these volumes feature a technological notion of communication while neglecting other forms of urban communication subject to regulation. This volume seeks to place communication in the foreground of the examination of urban/municipal regulation.

Of course the reach and scope of government regulation are always subject to debate, but those questions of reach and scope are of even greater significance and complexity when framed in a First Amendment context. The First Amendment appears on its face to prohibit government restrictions of speech, press, and the right to assemble, all issues germane to urban communication regulation. The First Amendment is always the elephant in the room, rightly so, we believe, when considering urban communication regulation.

It is with that broad perspective, that communication, broadly conceived, matters to the quality of life of individuals and so the broader society—both in terms of the aesthetics of individuals and the societies, and in terms of the ability to grow and govern smartly, and that communication in urban areas is profoundly important, because that is where most people are and where potential conflicts and opportunities are magnified by the density, and finally that government regulation or policy that affects communication is present and important, that we share the insights and stories of thirteen scholars and thinkers.

The intent of this volume is not to be overly "New York City centric" but to use this city as a prototype revealing how regulations can shape the communication landscape of any city. We have grouped the essays in this volume into three foci. In the first section, the authors look at content-centric issues in urban communication regulation.

In Chapter 1, Bauman examines the history of public art in New York City. And by "public art," she means government-sponsored and supported art. Generally driven by the gentry, there has been a long history of the City directing funds and other kinds of support and encouragement to art that helps set the public tone of the city. Art is sometimes seen as less than instrumental. Perhaps it is a luxury. Yet, to people living in crowded urban areas, to people coming from old worlds to new, art can play pivotal roles in helping them understand their community, its values, and its connections to others. It can sooth, inspire, excite, as well as agitate, mystify, and offend. Should the government, the same one that is supposedly barred from abridging speech, favor some expression over others? Should the government support some arts and not others? And if so, how? Clearly public art is chosen because it communicates. The intersection of government policies that promote or affect public art is the focus of this chapter.

Chapter 2 looks not at government-supported public art, but rather at the importance of public architecture and the municipality's attempts to shape communication through its buildings. Yes, as Faith Rose begins her chapter, "public buildings convey cultural and political meaning." Who has not felt walled out and left cold by ugly plain massive boxlike structures? And who has not felt invited in by warm, decorative, transparent, human-scaled structures? Design matters, and architect's vision communicates. Rose, the recent Executive Director of New York City's Public Design Commission shares examples where city building design—of libraries, firehouses, and the like—failed or succeeded in connecting with their communities and sent the "right" message. And indeed, there are messages to be sent, and architectural design sends them. Sometimes well, sometimes not. The design of public buildings may be left to individual architects or to individual departments that are paying for the construction. Alternatively, cities may take a broader and more proactive approach to designing their structures. Rose argues that cities' buildings can communicate and instill pride and encourage healthy citizenry through careful and thoughtful design that is administered by the city government. Public art and public architecture may represent the opposite of graffiti, communication and art sprayed onto the public's awareness which is at the heart of Gene Burd's analysis

In Chapter 3, Gene Burd takes on the issues of urban communication regulation with a frontal assault. He examines the robust and often controversial application of graffiti in cities. This frequently maligned form of communication is often attributed to the lowest elements of the urban poor and perhaps lawless. It may too reflect clashing values of young and old. Its messages are large, bright, colorful and designed to get into the faces of everyone around. They sometimes use code that may make the mainstream feel like uncomfortable outsiders. And graffiti artists/communicators/authors generally use others' property as their canvass. Potentially offensive in-your-face messages that "ruin" others' property is the very nature of graffiti and has a long history, as do efforts to put an end to, or somehow curtail such communication. By necessity, regulations designed to curb or limit graffiti deal with questions over whether such expressions are art, obscene, political, or commercial expression, how they might be protected by the First Amendment, as well as issues related to trespassing and personal or municipal property rights. What makes this all the more interesting is the ubiquity of graffiti. This is not a new phenomenon, nor is it uniquely American. At a moment when our gazes are increasingly focused on screens, graffiti's attempt to break through the clutter is all the more important, as are the attempts to channel or curtail it.

One might argue that the communicative function of graffiti, public art, and public buildings is not always central. There is no such argument that could be made regarding signs. Signs are designed first and foremost to communicate. Indeed, they generally are designed to be intrusive enough to communicate through the clutter that cacophonous cities offer.

In Chapter 4, Harvey Jassem explores the nature of signs and cities' complex efforts to regulate them. Like graffiti, signs are designed to be seen. Signs that whisper rarely call out for regulation, though in the strange world of New York's Times Square, the quiet and tasteful sign may be less legal than the bright blinking billboard. But it is the intrusive nature of signs that generally makes them subject to regulation. And while it may be readily apparent why signs might be subject to government regulation, in terms of their content, size, lighting and colors, placement, and such, what is less clear is just what makes something a sign. This exploration sheds light on what aspects of signage may be regulated, why those regulations may differ from place to place, and the difficulties in even determining what is or is not a sign. Is a pile of old tires a pile of blight that the city should remove, or is it a sign for the tire store on whose sidewalk it stands? How can the city protect both the sanctity of home and the right to communicate intrusively?

All regulation and law is place based. United States law applies in the United States, state law applies in that state, and city law applies in that city. There are even smaller contexts carved out for some communication regulation. The second section of this volume examines the role that place plays in urban communication regulation. Each chapter in this section deals in some fashion with zoning, so we start the section with a revised classic piece that explores the history and purposes of municipal zoning.

Susan Drucker and Gary Gumpert wrote a widely-cited piece on urban zoning in 1991. It is republished here, revised and updated, as our Chapter 5. Zoning regulations are designed to manage the character of a city's development. Generally, they focus on both the built and the use environments. Hence, they might direct that noisy dirty factories be built and operated only in areas that do not abut residences where people could have a right to expect some degree of quiet and safety. The regulations are, of course, far more detailed than that, and while they focus on what some might deem as the minutia of building and usage (detailing such things as permitted color pallets, materials, heights and setbacks, hours of operation, lighting, etc.), Drucker and Gumpert note that "zoning laws regulate face-to-face interaction restricting or stimulating communicative activity either explicitly or implicitly." And that communicative focus with regard to zoning that makes this chapter so important.

Chapters 6 and 7 focus on what have become known as "free speech zones." Underlying any discussion of "free speech zones" must be a consideration of public freedom to communicate. Clearly, that First Amendment privileges freedom of speech and assembly. Those form the basis of our democracy—people freely inform one another and share perspectives as we work to have an informed citizenry and electorate. At the nation's founding, the most fundamental places for such interactions were the public square, the sidewalks, and the parks. And yet, as central to our democracy as these public interactions are, they have sometimes caused problems that bump up against other important rights. In the crowded urban atmosphere, those clashes have led cities to develop regulations that channel public speech in ways that minimize disruptions of other rights. This is clearly a contentious endeavor. As David Allen writes in Chapter 6, "The regulatory structure that shapes the American idea of expressive freedom is often at odds with its philosophical roots." "How," he asks, "can a society that prides itself on having the greatest expressive freedom in the world, ... justify confining those who engage in political dissent [to 'free speech zones']?" Allen examines the legal philosophies that have been used to justify cordoning off free speech and argues for zoning regulations that will do a better job protecting speech.

In Chapter 7, Juliet Dee moves the discussion from the general to the specific. She zeros in on the regulatory attempts to zone speech in what might just be the United States' most cacophonous neighborhood, by design, New York City's Times Square. Times Square represents how effectively zoning can encourage an area to develop in a particular way. It is noisy, it is filled with flashing bright signage, it is filled with people. While such a cacophonous place might not be all that conducive to reasoned serious discussion and debate, the First Amendment says nothing about that. It says we have the right to communicate and assemble, so why, of all places, should Times Square be targeted as a place to limit both communication and assembly? Further, how would a city fashion regulations to accomplish such a goal? Dee takes us though those questions in this chapter.

To this point, all of the books chapters have considered urban communication regulation in the United States, where the applicable regulations are those of the duly elected government. Donald Fishman takes us to another "place" in Chapter 8, the "privatopia." Privatopias are privately owned "communities" which have their own governing systems that regulate physical and social appearances and interactions. Often, they are gated communities or buildings and operated as condominiums or cooperatives. One of the things they offer residents is predictability. One of the things they offer cities is a tax paying entity which pretty much takes care of itself. Privatopias are a growing segment of urban development, and as such, their regulation impacts a growing number of urban (and suburban) residents. Fishman discusses the emergence and growth of privatopias and the challenges they pose to traditional free speech rights. This is particularly important given that one's home had traditionally been the place where one had the most expansive set of personal rights. Fishman looks for ways that free speech might continue to be protected for those living in privatopias.

Finally in this section, Kevin Carragee looks in Chapter 9 at zoning not so much for how it does or does not control communication environments in the city, but rather how zoning and the values of zoning were used as rhetorical devices by a group in Boston that was working to save an important urban space or resource. Although cities are filled with people and resources, and there are abundances of both, ironically, that fullness also creates some shortages of resources. In this Boston example, a local Archdiocese moved to convert a school to a space for administrative offices. For the community, the school served not simply as a place to send children to classes, but as a community anchor and center for community interaction. The threat of losing that communication resource led to the formation of a community group and that group's arguments relied on the specifics and philosophies of zoning. Zoning helped to shape the discourse around the dispute and ultimately to its resolution. This is an interesting example of a rather indirect way in which regulations can impact urban communication resources.

Most definitions or descriptions of communication note that at its core, communication involves transferring a message or meaning. That requires some sort of a communication channel and manner of transfer. The third section of the book looks at issues related to channels and manners of communication.

The communication issues that have been covered thus far have been of the sorts that could have been discussed for eons in the past. Graffiti, public art, public architecture, signage, zoning, public space and the like. More recently electronic channels of communication are playing significant roles in urban communication, and they raise of host of important issues that are somewhat unique. In Chapter 10, Charles M. Davidson and Michael J. Santorelli look at how cities do and should do in order to encourage urban broadband access. Electronic communication has always relied on some sort of radio or wire/fiber carrier as a channel, and those have long been the result of some sort of private/public relationship in the United States. The social value of communication networks is enhanced the more accessible the network is. But building a network that can reach everyone is a technical, physical and financial challenge that necessitates deep pockets and the use of public infrastructure. Reliance on private entrepreneurs to build the channels still invokes public issues—should there be limits on the number of wires overhead and underground, if so, who gets to use the conduits, should there be conditions attached to the opportunity to use public conduits/streets, should those limits involve pricing, content control, universal access, etc? Reliance on the municipality to build the infrastructure, for the information superhighway, as municipalities build roads and sidewalks, raises a host of other questions—do cities have the finances and expertise to build such systems, should governments be in the position of overseeing communication channels, etc? Davidson and Santorelli look at these issues as they suggest best practices for enhancement of urban broadband.

Emily Long Vito takes us one step further. Once that broadband superhighway is in place, then what? Unlike the street corner where one can speak up, hold up a sign, spray some graffiti, using broadband to communicate requires skills, finances, and equipment beyond anything previously required for communication. In Chapter 11, Long shares New York City public-private partnerships that helped make the channels more accessible and helped make more people ready and able to use them. Such partnerships have a great deal to offer and can be very helpful, but, as Long points out, some of the problems are not easily addressed. And while there is much to be learned from some of these innovative programs, they provide some valuable tools and not the entire toolbox.

In Chapter 12, Drucker and Gumpert take us back to the street, but with a different angle. Some regulations deal directly with communication. Loudspeaker volume limits, signage size or placement limits, free speech zones, clearly focus on communication. Some regulations, such as many zoning regulations, deal with communication as a secondary impact. But what about food-related regulations? Drucker and Gumpert give new insight on rules targeting where we can chew the fat, so to speak. Food can offer a variety of contexts that engage people in communication, and street food can especially offer communication contexts that may be different than those of a person's non-food life. Consider the interactions between customer and immigrant street food vendor. These people might never come into contact with each other outside of this encounter. They might start with a conversation explaining the food, the food culture attached to the food, the immigrant's experience "back home" and here in the new city. Other customers who are standing around might well join the conversation in ways that are less likely in a formal restaurant or if folks were just walking by and not standing around eating together. But the opportunities for folks to gather around street food are impacted by street food vending regulations. While those regulations might portend to be focused on public safety, traffic, preservation of local brick and mortar restaurants, and the like, Drucker and Gumpert focus their analysis of the regulatory environment on the impact they have on this special intersection of communication.

All communication is shaped by not only the intentions of the communicator and the interpretations of the receivers, but also by the context or environments through which the communication is sent. The urban regulatory web plays a significant role in shaping urban communication that often goes unnoticed. This volume addresses this consequential yet overlooked connection.

References

Drucker, S., & Gumpert, G. (1996). The regulation of public social life: Communication law revisited. *Communication Quarterly, 9*, 7–32.

Georgiou, M. (2013). *Media and the city: Cosmopolitanism and difference.* Cambridge, U.K.

Johnson, N. (1972). *Test pattern for living.* New York, NY: Bantam Books.

Kohn, M. (2004). *Brave new neighborhoods: The privatization of public.* London: Routledge.

Robertson, A. (2010). *Mediated cosmopolitanism: The world of television news.* Cambridge, U.K.

Shiffman, R., et al. (2012). *Beyond Zuccotti Park: Freedom of assembly and the occupation of public space*. New York, NY: New Village Press.

Townsend, A. (2013). *Smart cities: Big data, civic hackers, and the quest for a new utopia*. New York: W. W. Norton & Company.

Tribe, L. (1978). *American constitutional law* (University Treatise Series) (3rd ed.). Foundation Press.

Zick, T. (2008). *Speech out of doors: Preserving first amendment liberties in public places*. New York, NY: Cambridge University Press.

PART ONE

Content and Regulation

CHAPTER ONE

Didactic to Collaborative

A History of Public Art Policy in New York City

EMILY BAUMAN
Institute of Fine Arts, New York University

In 1856, Henry Kirke Brown's equestrian statue of George Washington was installed in Union Square. Thousands of people gathered to mark the dedication of New York City's first public sculpture in a municipal park. Paid for by donations from private citizens, the monument was regaled by *The New York Herald* as an achievement by the art-loving city residents and "the commencement of the good work of beautifying our great metropolis …" (Kuhn, 2007). The following years proved this prediction as the city and its residents clamored to enhance their parks and buildings with statuary and monuments in an effort to promote a civic ideal and cultural status. This ushered in a legacy of collaboration between the public and private sectors to support the proliferation of public art that exists to this day. Public art now resides throughout the city's parks, on street corners and along the medians of its major avenues, in the plazas of corporate structures, and within its federal buildings.

Over the years definitions of both the terms "public" and "art" have changed drastically, suggesting roles for public art ranging from edifying and aesthetic purposes to transformative and democratic ones. While its origins in New York and in the United States in general lie in an invested didactic effort by those with wealth and power, public art has emerged in the second half of the 20th century as a practice that is both supported by and attempts to contend with this power structure. Postmodern thought has called for a reclamation of the public sphere for its daily user, a return to the notion of the commons, "the stage where the predictable and unexpected theater of the public could be presented and interpreted," where dissent and negotiation can openly occur (Phillips, 1989, pp. 332–333). Literature on public art programs, from the privately sponsored and non-profit organized to the federally funded, repeatedly return to the rhetoric of aesthetic improvement, community outreach and civic enhancement. This fundamental belief in public art's ability to communicate with and transform its surrounding community exposes the ideological underpinnings that simultaneously keep the practice alive while constantly complicating the process, reception and success of public art. Therefore, a defined purpose or role for public art—who it speaks to and for, what message it should communicate, why it is there—is necessary in developing policy regarding the installation of art in the public sphere.

This chapter seeks to investigate the history of public art in New York City in order to delineate the genesis of current public art policies and the complex ideologies that inform current expectations for it, in order to work towards defining what makes for a "successful" public art program and lay out a roadmap for improving public art's effectiveness in the city. Through this history, it will become clear that the "success" of public art depends on enthusiastic cooperation between the government that controls the public sphere, the community that inhabits it and support from private citizens and organizations that fund it. All three of these groups must be taken into account and included in the conversation and planning in order for an artwork or artistic program to be accepted as a positive and transformative addition to a neighborhood or site. It is due to their collaboration that today New York is one of the nerve centers of the international contemporary art world with numerous resources, foundations and legislation established to support a vibrant public and private art scene.

Government organized agencies and programs, including Percent for Art legislation and Art in the Parks programming, along with the efforts of New York City's Department of Cultural Affairs and Department of Parks and Recreation,[1] among many others, contributed to this push for public art and the provision of sites for its display. Non-profits like Creative Time, Inc., the Public Art Fund, Inc., and Mad. Sq. Art, as well as prominent corporations from Philip Morris to UBS, commercial art galleries and museums have assisted in organizing, promoting and funding many of these ventures. The result has been to transform public space into an outdoor museum, exposing the public to art no longer tamed by the bounds of institutional walls (Heckscher, 1967). The genesis, organization and funding of programs such as Percent for Art and Art in the Parks serve as examples of promising prominent models of public art policy that can help other cities develop or improve their own public art programs.

So, how did New York transform from a cultural desert into a cultural mecca in less than one hundred and fifty years? The legislation and organizations listed above only date back, at the farthest, to the 1960s, when a sea change occurred in the attitude regarding federal support for the arts. This instigated a windfall of new programs and groups that rallied around the installation of art in the public sphere. Through these projects the New Frontier rhetoric of "American individualism" has been molded to match the postmodern commitment to diversity. But even this contemporary era of public art owes its due to the legacy of public art in the city that dates back to the Progressive Era, the cultural institutions that it created and the Modern Art museums that were established towards its end.

The history of art in New York, and in the United States at large, has been embroiled in debates surrounding public morality, democratic ideals, and revolution since the country's inception. It originated in a top-down didactic effort to communicate civic ideals, encourage unity and instill patriotism; a moralizing role for art that would also serve to assert the cultural legitimacy of the nascent republic. New York City became one of the major testing grounds for the efficacy of public art, with its "economic, political and cultural elites" creating a consolidated effort to make the city a beacon of culture to the rest of the expanding country (Bogart, 1989, p. 4). In doing so, an ideology was established that continues to inform people's expectations for public art to this day.

Didactic Democracy: The Origins of Public Art in the US and NYC

Colonial America was a land of merchants, farmers and tradespeople, whose limited wealth was tied up in trade, settling and expansion. It wasn't until the time of the War of Independence that the colony reached a point of wealth and stability to begin to consider cultivating the arts.[2] However, after the war the young nation was still reeling from the stifling incursions of colonial control. The great monuments of Europe, from obelisks and triumphal arches to cathedrals and castles, commemorated wars, empires, monarchs and the Church. Placed in commanding positions at the apex of boulevards and in the middle of public squares, their heroic imagery communicated grand narratives of wealth and power, as well as ones of subjugation and control. Having thrown off these decadent vestiges of power, the country faced a crisis regarding both the proper role of art in a democratic society and appropriateness of federal patronage of the arts. Many major figures, including John Adams, argued against a place for art in the United States, regarding it as the product of despotic regimes, unfit for the new republic (Miller, 1966, pp. 3–13).

Lillian B. Miller (1966) explained that, despite these concerns, the will to celebrate its new political system and seemingly unstoppable Manifest Destiny, as well as prove its cultural equality (if not superiority) to Europe, won out. Strongly influenced by 18th century Enlightenment ideals and British neoclassicism—distributed through the writings of Sir Joshua Reynold, Archibald Alison and William Wordsworth—art was seen as both a moral tool in the effort to improve society and a way to develop Taste and Rationality. The following generation of Romantics—Ralph Waldo Emerson, John Ruskin and Charles Eliot Norton—directed artists, like the Hudson River School painters, towards nature as a subject, imbuing the unblemished landscapes being settled across the U.S. with a holy character that justified Manifest Destiny (Miller, 1966, pp. 1–32).

Nationalists, like New York's DeWitt Clinton, believed that images of "the truly great of the nation" and "deeds of patriotism and heroism" would preserve the country's unity, investing in a strong native economy and national culture supported by government and community patronage (Miller, 1966, pp. 10–22). With the privately funded installation of Brown's sculpture in Union Square and the creation of Central Park in the late 1850s, enthusiasm about landscape statuary grew rapidly in the City. Citizen groups began offering gifts of public statuary to their parks as ways to legitimize American heritage and beautify the city beginning c.1860 (Bogart, 1989, p. 18).

The Civil War (1861–65) shook the belief in America's moral superiority to its mother countries. In addition to creating a demand for commemorative monuments, the war increased the desire to create nationalistic images of unity that reasserted American values. In New York, Michele H. Bogart (1989) described how a "self-conscious movement" or "crusade" arose among the city's elite to instill a *civic ideal* by integrating sculpture and monuments into city planning to offer "an urban vision of patriotism, civilization, and good government." Among these patrons were the enormously wealthy robber barons (Carnegies, Rockefellers, Vanderbilts) who saw their role in society as "modern-day Medicis" in charge of recovering the country's spirit through the aesthetic enhancement of the urban environment (Bogart, 1989, pp. 26–27).

American artists returning from studying at the Ecole des Beaux-Arts produced monuments that echoed the dramatic grandeur of Haussmann's renovated Paris. Among them was John Quincy Adams Ward, a native New York sculptor, whose *Indian Hunter* (1869) was the first work installed in Central Park. Bolstered by France's donation of Frédéric Bartholdi's *Statue of Liberty* in the 1870s, these monuments sparked a landslide of other commissions throughout the park and the city. The massive journalistic and civilian campaign to fund a base for Bartholdi's statue led to the penning of Emma Lazarus's *New Colossus* and its adoption as a national motto.[3] The success of the domestic fundraising effort reinforced the national interest in supporting major public sculpture (Bogart, 1989, pp. 19–23), proving that art could communicate both a national ideology and identity.

Chicago's World's Columbian Exposition of 1892 was the lynchpin for the artistic movement. The outrageous success of the White City inspired the foundation of the City Beautiful movement: an umbrella approach to public beautification that espoused the municipal benefits and reform that came from good design and comprehensive city planning (Progressivism).

Figure 1.1. Looking West from Peristyle, Court of Honor and Grand Basin of the 1893 World's Columbian Exposition (Chicago, Illinois).

Like dominoes, a string of societies and organizations began to form in the 1890s to enact City Beautiful agendas, including the National Sculpture Society (NSS), Municipal Art Society of New York, and Fine Arts Federation of New York. These groups worked to enhance the professional acceptance of artists, promote exhibitions and commissions for public art, ensure artistic quality in public works, and embrace the broader causes of historic preservation and municipal planning. In turn, they helped push for the formation of broader, government-related organizations including the Art Commission of the City of New York (1898), which was written into the revised city charter after the passing of the Greater New York Legislation. The Commission consisted of a committee of appointed members from the art world and government (Bogart, 1989, pp. 26–66). These organizations worked together to set guidelines for the design of public fixtures, mark historic buildings for preservation, commission works for donation to the city, consult on the placement of public artworks, and more.

The combination of the Greater New York Legislation (1896), which consolidated the land that is now designated as the five boroughs into one city, and the Pendleton Act (1883), which expanded federal government activities, led to an influx of federal building construction in New York around the turn of the century, including the Hall of Records, Municipal Building, Customs House and major institutions like the NYPL, Metropolitan Museum and Brooklyn Institute. In the 1890s the NSS drew up guidelines for public monument competitions that have remained relatively unchanged: they assured anonymity, fixed costs, transparency about prizes, expectations and date of submission, as well as transparency about jurors with a promise to include experts on panels (Bogart, 1989, pp. 73–85).[4]

The result of the Beaux-Arts, City Beautiful and Progressive movements was a vast array of seemingly uncontroversial public sculptures that resembled European precedents. Sponsored by wealthy New Yorkers, sensitive to the underlying threats that social changes (depression, immigration, industrialization) posed to their way of life, the purpose of the art was to communicate the grand ideals of the upper class to the masses through allegories and myths relating to modern life, nationalism and American might (Bogart, 1989, pp. 221–232). However, along with these elitist sensibilities, the transition from the Victorian to Progressive era marked an increased interest in a more efficient, sanitary, scientific approach to architecture. The City Beautiful interest in overall design and municipal reform reflected an engagement in international discussions about social democracy leading to studies of diversity in population using documentation and statistical analysis, resulting in major changes in zoning and housing laws (Wright, 1983, pp. 158–177).

Changing of the Guard: Depression, World Wars, and a Decline in Public Art

Bogart (1989) described the emergence of a new generation of college-educated, business-minded men in the 1920s whose interests diverged from those of their genteel Progressive parents (pp. 293–294). Having seen and experienced the devastation of the Great War, the failure of elite moral mythologies seemed apparent. The aesthetic focus of those who promoted the City Beautiful movement was dwarfed by the pragmatic elements Progressivism had ushered in. Approaches to corporate structuring began to change in the 20s as many corporations consolidated into trusts. Beginning with the Gothic-style Woolworth Building (1913), the concept of the "Cathedral of Commerce" was introduced: gargantuan skyscrapers dedicated to celebrating the success of self-made men. The many Art

Deco skyscrapers that followed—Irving Trust (1929–31), Chanin (1929) and Chrysler (1929–30) Buildings, for example—offered new opportunities for architectural commissions, but for a new generation.

The new artists looked to the industrial age, using curved lines and simplified forms seen in machinery. The 1913 New York Armory Show had the same kind of revolutionary effect that the World's Columbian Exposition had on artistic style and outlook over the preceding two decades. It introduced Americans to Cubism, Fauvism and Dadaism, among other styles, changing the tide of art towards international, modernist, abstract and avant-garde.[5] These reflected the values and tastes of its new patrons from the business community, who consolidated into a powerful pro-avant-garde class whose prestige and influence came to eclipse that of the preceding City Beautiful Movement. Their efforts resulted in the creation of the Museum of Modern Art, the Whitney Museum of American Art and the Guggenheim Museum.[6]

During the following years of World Wars and the Great Depression fewer and smaller federal buildings were commissioned. The notable exception in NYC was the Roosevelt Memorial wing of the Natural History Museum (1936). With its triumphal arch entryway and equestrian statue frontispiece, it was one of the only state funded buildings commissioned during the 1930s and perhaps the death rattle of the City Beautiful movement. Meanwhile, a new model of civic building and beautification was initiated when John D. Rockefeller, Jr. introduced Rockefeller Center as the largest privately owned complex of business and entertainment, with over one hundred works by thirty artists integrated into its comprehensive design scheme. An art advisory committee of museum directors and scholars consulted with the architects to choose artists. Like City Beautiful proponents, they chose timeless allegorical themes about progress and stability including *Prometheus* and *Atlas*, but used stylized, streamlined imagery to reflect modernity, communication and entertainment: the industries that Rockefeller Center was built to host.

During the Depression New York City turned first to the state and then to the federal government for work relief. The city received a significant amount of money from the Works Progress Administration (WPA), with some of it coming directly from the Federal Art Project, which codified the inclusion of commissions of artworks as part of the renovation and creation of federal buildings and amenities.[7] Many public amenities were built, including the 79th Street Boat Basin (1937), Astoria Park Pool (1936), Prospect Park Zoo (1935), and the Bronx-Whitestone Bridge (1939) as a part of Robert Moses's Belt Parkway system around Brooklyn and Queens. Included in these plans were artworks, however, Moses had little interest in the arts beyond their use as decoration, leaving a legacy of pretty but uninteresting outdoor public art from the era. The direct funding of artists by the

WPA was an outlier in American history, one that is rarely repeated, but does reappear in the template for Percent for Arts legislation.

After World War II, federal arts programs kept failing. Talk of creating a federal arts commission sprang up repeatedly but died in committee. Cultural opponents continually railed against federal art support during the Truman and Eisenhower administrations "as a return to WPA boondoggling, a meddling threat to the 'rugged individualism' of artists ...", while right-wing congressmen like George A. Dondero (1949) derided "brainwashed artists in the uniform of the Red art brigade," even before McCarthyism began. Despite the formation of The Committee on Government and Art in the late 1940s to lobby for arts policies, the government's Commission of Fine Arts declared the government's involvement in arts funding sufficient in a 1953 self-assessment (Wetenhall, 1989, p. 303). Partially this reflected a turn away from decoration in design as the International style of architecture gained in popularity[8] and its slick glass box skyscrapers became the signature for corporate America, introduced by NY's iconic Lever House (1951) and Seagram's Building (1958).

Additionally, the lack of money going into the arts at home was due to the focus on promoting American art abroad during the Cold War. Nancy Jachec (2002) delineated how Abstract Expressionism was advanced in Europe (1958–61) by the State Department in order to establish an "Atlanticist, or Euro-American aesthetic" that could woo unaligned intelligentsia and factions in Western Europe towards American aims and interests (p. 105). The deliberate propagandistic intent of this plan was exposed by the abrupt cultural policy changes in 1962 under the Kennedy administration and the 1967 CIA exposés. The Kennedy administration backed off the idea of a totalizing Atlanticist aesthetic in favor of promoting America's avant-garde as distinct from Europe's, leading to the conception we have today of Abstract Expressionism as an intrinsically American aesthetic (Jachec, 2002, pp. 105–106).[9]

Resurgence of Art: New Frontier, NEA and Art's Return to Architecture

John Wetenhall (1989) chronicled how Kennedy's presidency marked a significant change in policy regarding the arts. The enormously positive reaction the public and media had to inviting Robert Frost to speak at his inauguration led JFK to begin including cultural leaders at state occasions (p. 303). When the 1961 Metropolitan Opera labor impasse necessitated Secretary of Labor Arthur Goldberg to intervene at Kennedy's behest, Goldberg's written decision served as the first federal policy statement since 1953 and established precedent for federal participation in the arts. He identified the crisis as the result of rapid cultural growth whose demands were exceeding the limits of private philanthropy. As a solution, he proposed that the government match local grants, offer tax incentives for private donations and establish a council to study additional possibilities. "In sum," Wetenhall (1989) explains, "Goldberg advocated a new federal responsibility equal to that of health, education, and welfare," establishing support of the arts as integral to public policy and economic growth (pp. 303–304).

The success of the mediation led to the creation of the Ad Hoc Committee on Government Office Space, whose report "Guiding Principles for Federal Architecture" (1961), pushed for better, innovative design for public structures. A policy requiring the government to spend up to one percent of a federal building's cost on art was omitted because it was instated separately before the report's publication. Percent for Art legislation lapsed in 1966 due to the Vietnam War, but was revived in 1973 (Wetenhall, 1989, p. 304), and now accounts for a large amount of art in public buildings, especially after its expansion to the state level.

Subsequently, August Heckscher was brought in as an outside consultant to conduct a federal "inventory" of cultural activities. His massive report, *The Arts and the National Government* (1963), endorsed percent for art legislation, the use of independent panels of experts for aesthetic consulting and design and the creation of an Advisory Council and Foundation to be organized like the extant ones for health and science. Through Heckscher's guidance, Kennedy put into place a number of experimental programs, many of which later became public policies. One week before the report's publication, JFK issued an executive order creating an Advisory Council on the Arts. Although it was delayed until after his assassination, President Johnson appointed a council and signed legislation creating the National Endowment for the Arts (NEA) in September 1965, establishing what Wetenhall (1989) concluded is "the ideological foundation for the government's active cultural role that persists to this day" (pp. 305–306).

The NEA adopted the Art in Public Places program in 1967 and helped revise the General Services Administration's (GSA) Art in Architecture policy in 1973. Art in Public Places became one of the most controversial and visible programs of the NEA, funding over seven hundred works across the country between 1967 and 1995 (when the program ended due to massive budget cuts). Unlike Art in Architecture and other federally sponsored projects, Art in Public Places offers to match up to $45,000 for community-initiated projects (Beardsley, 1981, pp. 10–13).

According to John Beardsley (1981), the program "assists communities across America in their efforts to increase public awareness of contemporary art. It seeks to support the enhancement of the public environment through art …", resulting in artworks that "are expressions of a community's aspirations for itself." Applications required information about the selected artist, the proposed site, a description of the selection process, evidence of support by the community and civic authorities, and a plan for community outreach (Beardsley, 1981, pp. 10–13).[10] However, despite the procedural inclusion of the community the ideology Beardsley espoused often fell short. Rosalyn Deutsche (1988) explained that his inclusive rhetoric hides that these collaborations were often with reformers whose impetus was gentrification of minority and working-class neighborhoods through beautification campaigns that claimed to boost local morale, but instead imposed control and order on public space (pp. 19, 29). Although sponsors ranged from ad-hoc citizen committees to existing non-profit organizations and arts institutions, the problem of representation of the community remained an issue, as they were rarely the actual instigators of said projects.

Similar problems were encountered when Mayor Koch instituted Percent for Art legislation in New York City in 1982. At the urging of Deputy Mayor Ronay Menschel and Doris C. Freedman he passed the law that *up to* one percent of a new public building's budget must be devoted to art—the cost cannot exceed $400,000 total, unless the City agency decides to waive the cap. The decision to implement this law occurred at a key moment of public awareness regarding public art due to the intense controversy surrounding the installation, and eventual removal, of Richard Serra's *Tilted Arc* sculpture in Foley Federal Plaza in downtown Manhattan. An extreme example of a failure in public art execution, the Art in Architecture commission of Serra's work in 1979 was proposed and instituted without any community outreach or discussion. The work was met with an outraged petition signed by nearby workers, articles and letters to the editor in local and national newspapers, and culminated in public hearings held in March 1985. The large metal sculpture, which acted like a curved wall in the middle of the plaza interrupting ambulation, was regarded as an eyesore and inconvenience. The panel vote from the hearings determined that the sculpture should be relocated. Serra responded with a lawsuit, which was dismissed in 1987, and an appeal, but the government went ahead and removed the work overnight on March 15, 1989 (Kammen, 2006, pp. 238–246).[11]

This episode cast into high relief the questions "what is the purpose of public art?" and "who it is meant to serve?" The late 1980s was a particularly controversial time for the arts. Senator Jesse Helms and pro-censorship members of Congress openly waged war against government agencies that funded any art that might be considered offensive, leading to controversies surrounding NEA sponsored exhibitions and performances by Robert Mapplethorpe, Andreas Serrano and Karen Finlay. Resulting litigation against the NEA would continue until 1998 when the United States Supreme Court ruled on the issue. Meanwhile, the GSA changed its Art in Architecture process to include input by local communities during the nomination and selection process, halting twenty-five forthcoming projects pending the finalization of its new procedural framework (Kammen, 2006, pp. 243, 286).

Low Economy, High Art: Art in the Parks, Non-profits and Alternative Spaces

The crux of the issues that surfaced under the resurgence of government supported art recalls Beaux-Arts attitudes towards public art: that it was put in place by elites to influence, educate or control the public. The rhetoric of elitism continued in the 20th century in the grand moral claims made by art critics like Clement Greenberg and Michael Fried. Greenberg's introduction of the concept of "art for art's sake" (1939) dismissed the radical leanings of modernist artists as a means to an end, instead placing the avant-garde in an intellectual bubble immune to the influences of easy pleasure (kitsch).[12] However, the introduction of postmodern theory and its widespread discernment throughout the art world—especially through feminist and activist art—began to break down this hierarchical structure regarding public art starting in the 1960s. The art scene in NYC started to find ways to present art to the public in an experimental and communicative way, with Art in the Park programming as an instructive example.

New York's Department of Parks, Recreation, and Cultural Affairs inaugurated Art in the Parks in January 1967. The first major group exhibition, "Sculpture in Environment," opened in October in nine city parks and fifteen public and corporate buildings and plazas throughout the city. Freedman organized the show while she was working for its sponsor, the Office of Cultural Affairs. The exhibition included many of the most important artists of the time—Alexander Calder, Marisol, Louise Nevelson, Barnett Newman, Claes Oldenburg, David Smith—with more than half of the twenty-four artists in the show creating site-specific works for the City (Kuhn, 2007). The success of the first exhibition—including the adoption of Bernard Rosenthal's *Alamo* as a permanent installation at Astor Place, by popular demand ("Astor Place," 1967)—led to further collaborations with arts organizations and artists to bring a continuing series of temporary art installations to the city's parks.

Art in the Parks' formal application process for exhibitions relies on the city's legacy of collaboration between private funding and government controlled public land, requesting full proposals be submitted for approval by a departmental panel composed of the NYC Parks Commissioner or their representative and the local community board. Once the license is issued, responsibility for all funding, insurance and promotion of the work falls on the artist or sponsor.[13] Two important non-profits, the Public Art Fund and Creative Time (founded 1977 and 1974, respectively), are among the most active independent groups working with the Parks Department. Emerging during the city's recession, both of these non-profits responded to the financial burden put on the artists for outdoor exhibitions by turning to private and corporate donations to create a variety of experimental and temporary public art programs.

New York narrowly avoided bankruptcy in 1976, many of its neighborhoods and parks were falling into disrepair or being cleared out in order to rebuild and gentrify. Throughout the decade economic and social debates about race, sexuality and class were everywhere. Along with the non-profit arts organizations, New York was experiencing a burgeoning "alternative-space movement" of artist-run programs with a populist bent that were engaged with political critique. Drawing from recent artistic precedents—Conceptual, Minimal and performance art, etc.—they created socially engaged activist practices that blurred the roles of artist, curator and collaborator. By either bringing the city into the gallery or going out into the street, these artists used institutional critique as a way to engage with the urban landscape and local authorities.

The irony behind this critique was that government funding for the arts, from the NEA and New York State Council on the Arts, among others, was at a high.[14] These funds made it possible not only for artists to support themselves and for small non-profit spaces to crop up across the city, but for socially motivated interventions to be funded using government money. For instance, the government-subsidized spaces Colab and ABC No Rio launched "The Real Estate Show" by unlawfully occupying an abandoned city-owned storefront on Delancey Street and filling it with art and performances that addressed gentrification and profiteering. The artists were evicted by the police three days later at the bidding of the Department of Housing Preservation and Development (Thompson, 2010).

Public art in New York City during the late 1970s became a confluence of government sponsored, non-profit organized and guerilla installations. Walls were blanketed with punk rock posters, advertisements and informational signs, embellished with graffiti tags. Within this visual confusion, the medium of postering emerged as an effective means by which artists could (illegally) become visible within official space and critique institutional issues. Most notoriously Christy Rupp's *Rat Patrol* project, a response to the three week long garbage strike (May 1979) gained infamy after a woman was attacked by a pack of rats in the Financial District. The media frenzy that followed also caught Rupp's posters of a huge rat reproduced from the Health Department's files, pasted to the temporary walls of the vacant lot adjoining the site of the attack where garbage had piled up creating unsafe conditions in which rats thrived (Crimp, 1984, pp. 52–53). Rupp's appropriation of the Health Department's own image worked to confound radical guerilla art with government-sanctioned signage.[15]

At the same time, non-profits were also working *with* the government to fund artworks they hoped would improve the city. From 1978 to 1985 Creative Time utilized the vacant areas of the downtown waterfront landfill for its annual group show "Art on the Beach" (Kuhn, 2007). As part of the Art in the Parks program, Battery Park City Authority, a state-funded public benefit corporation developing the land, donated the space. Each summer, the curators and artists contended with the new construction, developing socially conscious programs that would respond to the bulldozers and high-rises that surrounded their installations and performances (Contini, 1984, pp. 42–45). Although Battery Park City began as a housing initiative that would include a significant amount of low to middle-income residences, to address the city's housing crisis, it quickly turned into a series of luxury apartment complexes. By 1985, according to Phillips (1989), the aims of "Art on the Beach" artists had diverged from the aims of the developers and they stopped offering space to Creative Time. In 1987 the exhibitions resumed on a new site in Hunters Point, Queens owned by the Port Authority (p. 333).

Other groups working with Art in Public Places managed to get City employees and private companies excited about and involved in the creation of artworks. The Dia Art Foundation used Conrail's union employees to install Dan Flavin's fluorescent tubes on Grand Central Station's platforms (1976–86), and donated the work to the Metropolitan Transit Authority (while Dia continued to maintain it) (Beardsley, 1981, pp. 48–51).[16] Creative Time coordinated a massive effort to get New York City Transit Authority and Metropolitan Transportation Authority, as well as private companies including 3M, Westinghouse and Lighting Unlimited, involved with Bill Brand's *Masstransitscope* (1980) in the abandoned Myrtle Avenue subway stop (Contini, 1984, pp. 41–42).

So, how do these programs and organizations form and persist? There have been several efforts to designate given parks or areas within a park as permanent outdoor sculpture gardens for rotating exhibits under Art in the Parks programming. A number of these failed, including Dag Hammarskjöld Plaza (1971), Cadman Plaza (1972), Flushing Meadows Corona Park (1980s) and Hunts Point (Crames Square, 1998). But those that were able to obtain significant community support and form their own non-profit organizations with substantial financial support include Socrates Sculpture Park, which was successfully salvaged by sculptor Mark di Suvero in 1986 from a trash-strewn lot, Wave Hill, Snug Harbor Cultural Center and Mad. Sq. Art (Kuhn, 2007).

Madison Square Park offers a prime example of how community support and local funding contribute to the organization's survival, as well as the success and failure of individual installations. Despite its storied history and fortuitous location adjacent to the original Madison Square Garden, Flatiron Building and Metropolitan Life Insurance Buildings, Madison Square Park had fallen into disrepair by the 1990s. In 1994 it was graced by its first Art in Parks installation, a small fenced-in cornfield called *Circle of Hope* by Melinda Hunt, placed in the middle of the cracking and crumbling park to commemorate the anonymous dead of the potter's field that was once there (Kuhn, 2007). Supported by the City Parks Foundation, the work helped bring attention to the park and garnered interest for their Campaign for the New Madison Square Park, the precursor organization to the Madison Square Park Conservancy. They managed to raise funds for the renovation from companies operating nearby, including Metropolitan Life Insurance Company, New York Life Insurance Company, Credit Suisse First Boston, Rudin Management and Union Square Hospitality Group—many of which continue to support the park's Conservancy and its art program, Mad. Sq. Art, to this day.[17]

So, it would seem that "success" in public art has more to do with circumstance than great art, like, for Mad. Sq. Art, the willingness of its corporate neighbors to continue to invest in their surroundings. The same is true for individual installations, which rely on both funding and acceptance by the community. Teresita Fernandez's 2015–16 mirrored pergolas over the pathways of Madison Square Park invoked outrage in the neighborhood because it coincided with renovations to Shake Shack's restaurant stand in the park. The confluence of construction—with Fernandez's work looking like scaffolding during installation—upset residents who identify the park as a refuge of peace and quiet in the busy city (Kinsella, 2015). However, according to Miranda Siegel (2015) the brightly flashing lights of Rafael Lozano-Hemmer's massive *Pulse Park* in 2008–2009 received only positive responses. Because commercial buildings surround the park, the seemingly obtrusive ring of lights did not upset its neighbors, interfering minimally with the evening activities of nearby residents. Additionally, the work embraced park visitors through interaction, recording their heartbeats as the rhythm of the light pulses, "making the beating heart of New York visible" (Siegel, 2015).

Economic Art Boom: Percent for Art and the Precedent of the 1980s

The economic boom of the 1980s brought with it a return to prominence of the commercial gallery and a reinvigoration of art institutions. SoHo and East Village commercial galleries began cherry picking artists from among the non-profit groups. Federal funds began to dry out, the "alternative-space movement" drew to a close and curators and administrators from the non-profit sector began opening commercial spaces. The whole scene started to trend towards larger, more expensive, publicly commissioned works and exhibitions. For instance, the Public Art Fund's "Messages to the Public" series offered many up-and-coming artists who had been part of the alternative-space movement or graffiti culture, including Keith Haring, David Hammons and Nancy Spero a chance to show their work on the giant Spectacolor board at the center of Times Square. Jenny Holzer, who was involved in "The Times Square Show" (1980), which harshly critiqued the "Disneyfication" of the area, embraced the technologized Times Square by showing a selection of texts on the board for the entire month of March 1982 (Dickson, 2012; Thompson, 2010).

It was at this time of gentrification that Koch introduced Percent for Art legislation in NYC. With the success of the program in other cities,[18] as well as the numerous successful local precedents offered by installations by the Public Art Fund, Creative Time and Art in the Parks, Menschel recalls that the program was an easy sell to the City Council. "We made the case that engaging artists in the design of new facilities would contribute to City life and its public spaces." Since the Department of Cultural Affairs had the largest budget, they enthusiastically signed the law (Heartney, 2005, p. 30).

Percent for Art offered an alternative insertion for art in the city because its commissions are permanent and located in all new government buildings including schools, firehouses, police precincts, hospitals, juvenile halls, courthouses, subway stations, etc. Consequently they address constituencies ranging from city employees to students to criminals. The artists take on the role of a diplomat in these proceeding: working closely with the architects and engineers, relying on community feedback, integrating numerous requirements from the bureaucracies of various city agencies, and communicating with local press and politicians (Heartney, 2005, pp. 16–17).

The artists for these projects are chosen after the architect, and are often suggested by the architect from the program's slide registry or based on the suggestions of professionals on the staff or panel. The panel consists of representatives from the Department of Cultural Affairs, the agencies sponsoring the building, architecture and design firm employees and art professionals with ties to the borough or community. Constituencies that will be affected are invited to ask questions and express preferences during the first panel meetings convened to narrow down the list of artists. Then a second panel meets to discuss the finalist's presentations with non-voting observers invited including members of the community board, representatives of the Borough President, City Council and NYC's Art Commission (Heartney, 2005, p. 17).

The results of Percent for Art's projects have been mixed—some are beloved, some were calamitous and some artworks fade into the background. John Ahearn's experience in the South Bronx, despite his commitment to and communication with his neighbors there, exposes the flaws in the system's communication. His Percent for Art installation of three statues in front of the 44th Police Precinct were portraits of real people who lived in the neighborhood. *Raymond, Corey,* and *Daleesha* were intended as guardians for the neighborhood, in the tradition of the mile of heroes that stand guard at the Paseo de la Reforma in Mexico City. Although the work reflected his own admiration for and relationships with the downtrodden people in his neighborhood and his belief that the police needed to see them as individuals worthy of recognition, his images did not transform them, as he had intended, into heroes in the public's eyes. Instead they were read as racist stereotypes (Kramer, 1992). Unlike Kehinde Wiley's paintings of black men he meets on the street transported into the settings of Renaissance portraiture—an attempt to both reclaim a tradition of elevation and insert the black body into a history that denied it power[19]—Ahearn's statues failed to transcend. As Margarita Hunt-Tejada, District Manager for Community Board 4 in the Bronx, explained, what the neighborhood wanted "was an 'art' piece as opposed to a 'reality' piece" (Heartney, 2005).

The structure set up for public art in the 1980s continues to this day in New York with much of it still sponsored or subsidized by Percent for Art, Art in Parks, Creative Time, Public Art Fund, Dia Art Foundation and other arts-oriented organizations. Private funding that supports the non-profit organizations comes primarily from corporate donors and blue-chip art galleries. Most sponsor lists for outdoor exhibitions and museum shows contain company names including UBS, Morgan Stanley and Tishman Speyer, as well as the names of wealthy collectors and galleries or gallery owners. In February 2015 *The Art Newspaper* released their findings that "nearly one-third of the major solo exhibitions held in US museums between 2007 and 2013 featured artists represented by just five galleries" (Halperin, 2015). There is a certain extent to which this is inevitable: the top galleries have the most money and notoriety, therefore they can both attract and support the most successful artists, promising to extend their fame through the gallery's influence at museums and ability to fund projects such as those run through Art in the Park, which require full funding be found by the artists or sponsor. For instance, the primary donors listed for each of the 2015 installations by Mad. Sq. Art were the galleries that represented each artist.

This means that the role of the young, emerging artist in the sanctioned public sphere and in the museum is narrow. But this has always been the case. However, non-profits can fill that gap. For instance, Holzer began as a guerilla posterer, never gaining permission to hang her art in the public until her *Living* series circa 1980–82 (Ferguson, 1986, p. 70). The opportunities created for her and others by the Public Art Fund and Creative Time took them out of alternative practices and into the mainstream. In fact, the artists represented by the top five galleries most frequently exhibited in museum shows include many of the names that graced Times Square's Spectacolor board in "Messages to the Public."

Lessons Learned: The Elusion/Illusion of Success, Defining Public Art's Role

Adam Gopnik of *The New Yorker* describes Percent for Art's legacy as a "decorative, or intimate, conceptualism"; artworks that take up the ideas of Conceptual and Installation art and translate them for the public. For Gopnik these works function as a parallel to how the Rockefeller Center artworks reworked Cubism and Futurism into a civic language, claiming, "Art doesn't need to be examined closely every day to offer a deep experience over time. Its being there at all matters a lot" (Heartney, 2005, pp. 9–13). But this concept of art as quiet communication—a whisper, rather than a shout or, even as opposed to the Beaux-Arts concept, as an oration—is anathema to the definition of art for many in the art world, for whom art should be impactful and transformative.

Deutsche (1988) critiqued Koch's interest in municipal art commissions, planning and public art as resurrecting City Beautiful language by focusing on pleasant space and the eradication of unwanted uses of space (pp. 9–10, 17–18). For example, Koch focused on Union Square's history as part of the wealthy Ladies Mile district during the 1840s–90s (instead of its revolutionary past), to support his revitalization effort for the area. His flowery rhetoric camouflaged that the renovations specifically targeted the removal of derelicts in the park and the SROs surrounding it (Deutsche, 1998, pp. 6, 29).

As Phillips (1989) concisely explains, "Public art has been too often applied as a modest antidote or a grand solution, rather than perceived as a forum for investigation, articulation, and constructive reappraisal" (p. 331). Although public art find its genesis in the idea of the commons, a place of open communication and negotiation, this notion of the commons is in some ways a construct of bourgeois society, a false delineation between public and private that is encoded with the rules, ownership and controls that dictate the built environment. Deutsche's (1988) critique of revitalization shines a light on the power structures that underlie the presentation of public art (pp. 11–12).

With all of these conflicting notions about the role public art should play, what actually defines successful public art and how can it be instituted? Although Deutsche (1988, 1998) and others are right to critique the lip service that is often given regarding "community outreach", the reality is that public art is only truly successful when the community accepts and embraces its right to exist. In other words, it communicates with the public in a way that not only opens them up to art and its language, but allows that art, whether it reflect their values or not, to persist in their midst.

If the "successes" we see in contemporary public art really are, as Gopnik asserts, translations of esoteric art concepts into lay terms as "an expansion of the[ir] meanings" (Heartney, 2005)—or, even conversely put by Deutsche (1988), as a bastardization of the radicalism of late 20th century avant-garde art (p. 21)—does that negate their success or power? Or does the presence and success of this more decorative, de-politicized public art insert the language into the public sphere that actually allows avant-garde art to more easily communicate with and be understood by the public? Could it function as the art world's ABCs or tutoring program, teaching the non-museumgoing audience the basic language of art so that museums, galleries, and progressive high art become more accessible?

The first step in developing a public art program must be to define its intended purpose. Who is the art for, and what role should it play in the city's communicative structure? Harriet Senie and Sally Webster (1989) identify the public as generally uninformed about art yet "potentially powerful and politically enfranchised" (p. 288). Temporary exhibitions of high end, although accessible, art in public spaces create a viable purpose for contemporary art in the public sphere. They expose the public to a different form of visual communication while also opening them to the transformative/powerful messages of guerilla and political artwork, as well as more sophisticated art in the gallery and museum.

Although the involvement of communities in the selection and display of art may currently be flawed, Percent for Art and Art in Parks offer two models to be built upon. If the goal is to acknowledge the right of the lower class or disenfranchised to participate in the production of their own space—as opposed to leaving it solely in the hands of the wealthy and the government—then the processes and panels that are being set up by many of the government's public art programs and affiliated non-profits are moving in the right direction. Although Ahearn's sculptures represent a breakdown in communication, his insistence that they be removed because of that failure shows an acknowledgement of the public's right to control their own space. Additionally, Cai Guo-Qiang's 2005 work in front of the Bronx Criminal Courthouse, reflects this same consideration. The artist revised the piece after community feedback to his proposal identified his design for a chain of benches as reminiscent of chain gangs, a painful and aggressive connotation linked to the criminal justice system for a majority black and Hispanic neighborhood (Heartney, 2005, p. 29).

If public art acts like a course in art education (or at least artistic linguistics) through community resourcing and communication then it no longer functions in the same didactic, top-down way that the proponents of City Beautiful intended. Instead, it becomes a dialog between the public and the art world, connecting the sometimes obscure theory and forms that inform politically and socially engaged art back to the community who just might need that voice to shout or orate for it in a monumental form. If success in public art means that it is moving towards monuments and structures that better represent the needs and voices of the socially disenfranchised or the politically progressive, then art (whose aims have almost always been at the forefront of progress, often venturing into radical) is successfully being produced and incorporated into the public sphere.

Notes

1. The New York City Department of Parks and Recreation was called the New York City Administration of Recreation and Cultural Affairs when the effort to create temporary exhibitions in the parks in 1967 first began.
2. The Dutch in New York and the English in Boston and Virginia were minor exceptions to this rule. Taking up the mantel of European aristocracy, wealthy individuals in these areas began acquiring prints and commissioning portraits, with the earliest surviving paintings produced in the country dating from 1644. However, they had to rely on the work of artisans and craftsmen (most often local limners). Amateur painters and second-rate professionals coming from Europe – such as John Smibert – attempting to imitate European courtly tradition, produced awkward, stiff, ill-proportioned images of their patrons. See Bogart, 1989 and Miller, 1966 for more in depth discussions of the early development of art in the colonies.
3. For an in depth look at the funding and effects of the *Statue of Liberty* see Marvin Trachtenberg, *The Statue of Liberty* (New York: Viking Press, 1976).
4. The process of choosing an artist generally included establishing a committee (usually associated with one of the aforementioned organizations), raising the funds either from within the organization or on a wider scale, and selecting an artist through an open or limited competition (with city officials invited on the panel). Bogart, *Public Sculpture and the Civic Ideal*, pp. 73–85.
5. The Armory Show was organized by the Association of American Painters and Sculptors, a group of artists who broke off from the conservative National Academy of Design. In its formal statement, the AAPS said: "We wish to present at first hand the work of our artists, and that of artists in foreign countries which best give this permanent valuation of the awakening of a spontaneous individuality to a new use of art forms." The AAPS basically began and ended with the Armory Show. For more on the AAPS and the show see Milton W. Brown, *The Story of the Armory Show* (New York: Abbeville Press Publishers, 1988).
6. Wives of the business leaders lead the charge, with Abby Aldrich Rockefeller, Lillie P. Bliss and Mary Quinn Sullivan founding the MoMA in 1929 and Gertrude Vanderbilt Whitney establishing the Whitney Studio Club in 1918, which would become the Whitney Museum in 1931. The Solomon R. Guggenheim Foundation first opened its New York museum space under the name the Museum of Non-Objective Painting in 1939.
7. For a succinct overview of the WPA and public art, see Marlene Park and Gerald E. Markowitz, "New Deal for Public Art," in *Critical Issues in Public Art: Content, Context and Controversy*, Harriet F. Senie and Sally Webster, eds. (Washington: Smithsonian Institution Press, 1992): 128–141.
8. International Style architecture was named and defined by MoMA's 1932 exhibition on Modern Architecture. Wright (2008) concisely described it as favoring volume over mass, regularity without symmetry and no ornamentation (pp. 79–111).

9. For in depth analyses of the role Abstract Expressionism played in politics during the Cold War see, Jachec, 2002, and Eva Cockroft, "Abstract Expressionism, Weapon of the Cold War," in *Art in Modern Culture: An Anthology of Critical Texts*, Francis Frascina and Jonathan Harris, eds. (Phaidon, 1992).
10. This process was put into place after a few failed ventures on the part of grant recipients inability to fulfill their commitments of fundraising, installation or local approval. In 1979 the NEA instituted a letter of intent preceding the final application, so that they could assist with the process via advisory panels (usually art professionals selected by the Endowment) (Beardsley, 1981, pp. 12–13).
11. Serra insisted that the work was site-specific and lost its meaning if moved. The government claimed the right to remove it based on ownership – having commissioned and paid for the work. This incident had a strong effect on contracts and copyright laws in the arts going forward. For a full account, see Casey Nelson Blake, "An Atmosphere of Effrontery: Richard Serra, *Tilted Arc,* and the Crisis of Public Art," in *The Power of Culture: Critical Essays in American History*, Richard Wightman Fox and T. J. Jackson Lears, eds. (Chicago, 1993): 247–289.
12. Referenced here, Greenberg's "Avant-Garde and Kitsch" was originally published in *Partisan Review* (Fall 1939), but all of his essays can by found in *Clement Greenberg: The Collected essays and Criticism, vol. 1–4.* (Chicago: University of Chicago Press, 1986). Also see, Michael Fried, *Art and Objecthood* (Chicago: University of Chicago Press, 1998).
13. The proposed manner of funding must be included in the budget section of the application. For a full list of guidelines and applications see NYC Parks official website's "Temporary Outdoor Public Art Guidelines," http://www.nycgovparks.org/art-and-antiquities/temporary-guidelines (accessed 2/11/2016).
14. According to David E. Little, in 1977 the NEA made $3.5 million available for New York artists and the State Council on the Arts had another $9 million to distribute. "Colab Takes a Piece, History Takes It Back: Collectivity and New York Alternative Spaces," *Art Journal,* Vol. 66, No. 1 (Spring, 2007): 63.
15. Postering would remain a viable method of intervention and activism throughout the 1980s, reappearing in ACT UP's protests of homophobia and discrimination against people with AIDS. Gran Fury, a group of graphic artists, created the infamous "Silence = Death" imagery of an inverted pink triangle on a black background that were "'virtually impossible to miss' … In one weekend alone, twenty-five ACT UP volunteers wheatpasted more than 1,500 flyers in downtown Manhattan and Brooklyn …" Tamar W. Carroll's article "Turn Anger, Fear, Grief into Action ACT UP New York," in *Mobilizing New York: AIDS, Antipoverty, and Feminist Activism* (Chapel Hill, NC: University of North Carolina Press, 2015): 131–161.
16. The work was on Tracks 18–19, 39–40, and 41–42 until it was deinstalled c. 1986, according to Flavin's CV on David Zwirner Gallery's website, http://www.davidzwirner.com/artists/dan-flavin/exhibitions/permanent/
17. See Madison Square Park Conservancy's website for more information about its history and supporters: http://www.madisonsquarepark.org/
18. The first states and cities to institute Percent for Art programs include Hawaii, Washington, Colorado, Nebraska, Florida, New Jersey, Connecticut, Maine, Philadelphia,

Baltimore, Atlanta, Chicago, Kansas City, Tulsa, Seattle, San Francisco and Anchorage (Beardsley, 1981, pp. 10–13).

19. For more on Kehinde Wiley's work see *Kehinde Wiley: A New Republic*, Eugenie Tsai, ed. (Brooklyn, NY: Brooklyn Museum, 2015).

References

Beardsley, J. (1981). *Art in public places: A survey of community-sponsored projects supported by the national endowment for the arts*. (A. L. Harney, Ed.). Washington, DC: Partners for Livable Places.

Bogart, M. H. (1989). *Public sculpture and the civic ideal in New York City, 1890–1930*. Chicago, IL: The University of Chicago Press.

Cockroft, E. (1992). Abstract expressionism, weapon of the cold war. In F. Frascina & J. Harris (Eds.), *Art in modern culture: An anthology of critical texts* (pp. 82–90). London: Phaidon.

Contini, A. (1984). Alternative sites and uncommon collaborations: The story of creative time. In S. Paleologos Harris (Ed.), *Insights/on sites: Perspectives on art in public places* (pp. 39–47). Washington, DC: Partners for Livable Places.

Crimp, D. (1984). The art of exhibition. *October*, *30* (Autumn).

Deutsche, R. (1988). Uneven development: Public art in New York City. *October*, *47* (Winter).

Deutsche, R. (1998). Krzysztof Wodiczko's *Homeless Projection* and the site of urban 'Revitalization'. In R. Deutsche (Ed.), *Evictions: Art and spatial politics*. Cambridge, MA: The MIT Press.

Dickson, J. (2012). Recollections. *The Times Square Show—Revisited*, exhibition Sept. 14–Dec. 8, 2012. Retrieved August 10, 2016 from http://www.timessquareshowrevisited.com/accounts/jane-dickson.html

Dondero, R. G. A. (1949). Modern art shackled to communism, *Congressional Record-House*, 81st Congress. In P. Hills (Ed.), *Modern art in the USA: Issues and controversies of the 20th century* (pp. 190–92). Upper Saddle River, NJ: Prentice-Hall.

Ferguson, B. W. (1986). An interview with Jenny Holzer. In J. Simon (Ed.), *Jenny Holzer: Signs*. Des Moines: Des Moines Art Center.

Gopnik, A. (2005). Introduction: Art in the city. In E. Heartney (Ed.), *City art: New York's percent for art program*. New York, NY: Merrell.

Halperin, J. (2015, April 2). Almost one third of solo shows in US museums go to artists represented by five galleries. *The Art Newspaper*.

Heartney, E. (2005). The city as laboratory: Two decades of New York's percent for art program. In M. Heiferman (Ed.), *City art: New York's percent for art program* (pp. 15–23). New York, NY: Merrell.

Heckscher, A. (1967). Foreword. In New York City Administration of Recreation and Cultural Affairs (Ed.), *Sculpture in environment*. New York, NY: Seybert-Nicholas Print. Corp.

Jachec, N. (2002). 'A Partnership of Equals': Kennedy, the European Union and the end of abstract expressionism as an Atlanticist aesthetic (pp. 105–18). *Third Text*, *16*(2). Routledge.

Kammen, M. (2006). *Visual shock: A history of art controversies in American culture*. New York, NY: Alfred A. Knopf.

Kinsella, E. (2015). Parkgoers Furious over Teresita Fernández Sculpture at Madison Square Park. *Artnet*. Retrieved August 10, 2016 from https://news.artnet.com/exhibitions/controversy-teresita-fernandez-madison-square-287929

Kramer, J. (1992, December 21). Whose art is it? *The New Yorker*. Retrieved February 14, 2016 from http://www.newyorker.com/magazine/1992/12/21/whose-art-is-it

Kuhn, J. (Ed.). (2007). *The outdoor gallery: 40 years of public art in New York City parks*. New York, NY: City of New York Parks & Recreation, Arsenal Gallery.

Little, D. E. (2007, Spring). Colab takes a piece, history takes it back: Collectivity and New York alternative spaces. *Art Journal, 66*(1), 60–74.

Miller, L. B. (1966). *Patrons and patriotism: The encouragement of the fine arts in the United States, 1790–1860*. Chicago, IL: University of Chicago Press.

Phillips, P. C. (1989, Winter). Temporality and public art. *Critical Issues in Public Art. Arts Journal, 48*(4), 295–304.

Siegel, M. (2015). The annotated artwork: 'Pulse Park.' *New York Magazine*. Retrieved August 10, 2016 from http://nymag.com/arts/art/features/51352/

Thompson, M. (2010, Spring). The times square show. *Streetnotes* 20, *Urban Feel.*

Wetenhall, J. (1989, Winter). Camelot's legacy to public art: Aesthetic ideology in the new frontier. *Arts Journal, 48*(4), 303–08.

Wright, G. (1983). *Building the dream: A social history of housing in America*. Cambridge, MA: MIT Press.

Wright, G. (2008). *USA: Modern architectures in history*. London: Reaktion.

Preface to Chapter 2

Chapter 2 represents a unique and valuable contribution to the study of urban communication regulation and the roles played by local government in shaping civic messages. While most thinkers seek to shed light from the outside, Architect and former Executive Director of New York City's Public Design Commission, Faith Rose shares the observations of a quintessential insider. This essay provides readers the insights of a leader who was deeply involved with shaping New York City's architectural design decisions. Ms. Rose, who was awarded the American Institute of Architects New York State "Public Architect Award," shares, in this essay, the types of concerns city governments do and can address, and notes that given the number of stakeholders, shaping public design and the communication design affords, is complicated but critically important for healthy cities.

CHAPTER TWO

How Public Architecture Communicates

FAITH ROSE
Principal, O'Neill Rose Architects, AIA

Introduction

Public buildings convey cultural and political meaning. They communicate a city's level of investment in a neighborhood, as well as the degree to which the inhabitants are valued by city government. These messages are part of the environment of the city, which is inextricably linked to the health and wellbeing of its inhabitants.

A good architect is able to resolve myriad conflicting requirements within a singular, clear expression. This is the basis of good design, and is key to ensuring that buildings communicate their intended message. However, this talent is not enough to ensure our public buildings are well designed. The process of public capital improvement is lengthy and complex. Architects, landscape architects, and other design professionals cannot create successful public buildings and spaces without support from the governmental agencies tasked with overseeing public capital improvement.

Good public design is susceptible to many pressures. Other priorities, such as budget and schedule, are perceived to be in conflict with quality design. In addition, the changing goals of successive administrations, as well as the simple fact that the impact good design has on public projects is not widely understood, can jeopardize governmental programs created to support it.

Cities have a responsibility to support and advocate for well-designed public buildings and spaces. Civic spaces that purposefully communicate values such as equity and social resilience lead to a dynamic and just city. In addition, the most underserved portion of a city's populace use public buildings with greater frequency. This means that spaces which house public services can have a tremendous impact in bringing about equity and social resilience.

Environment cues behavior, and a thoughtful, humane environment goes a long way to creating a healthy and dynamic city. Public spaces within a city can promote such an environment, thus improving the quality of life of its inhabitants, which in turn affect a city's economic health. Given the pervasive lack of understanding that public spaces can have a positive impact on the health of a city, those responsible for public design must be able to clearly communicate its benefits.

This chapter will consider how buildings express meaning through an examination of New York City branch libraries. It will also explore how architects create buildings that communicate civic values through the development of the FDNY Rescue 2 station in Brooklyn. Finally, it will look at how municipal agencies and programs within New York City, specifically the Public Design Commission, the Department of Design and Construction, and the Design Excellence Program, support good public design.

How Public Buildings Communicate

Branch libraries and fire stations are found in most American cities, and are typically dispersed through neighborhoods rather than being centrally located. Therefore, they are very much "of the people," part of the narrative of cities and towns, and built into the residential fabric. As such, they are well situated to carry messages of a city's level of investment in its populace. The evolution of the New York City's library design over the last century can be used to demonstrate how buildings communicate cultural and political meaning, purposefully or otherwise.

A building is never mute; its presence may be more or less commanding, but it always carries a message. In the early 20th century, 67 branch libraries funded by Andrew Carnegie were built throughout the five boroughs of New York City. Like most civic architecture of the time, many of these branch libraries were based on the architecture of the Greek temple. Built with the noble goal of educating the masses, the Carnegie library communicates an aspirational message about the power of knowledge. A more subtle message, authority and power over these masses, is also conveyed.

For example, the typical Carnegie branch library does not meet the patron as an equal, on level ground. The user must ascend to the entrance, lifting themselves out of the common realm. The entrance doorway, over scaled and often flanked by tall columns, as can be seen in the Stone Avenue branch library in Brooklyn, is at once welcoming and overwhelming. Once inside, the clear delineation of spaces are another welcoming gesture. However, they are dominated by an imposing central desk that marks the librarian as an authority figure. The Carnegie libraries were for the people but were not of the people. Meant to inspire awe, their architecture diminishes the library goer, who is literally required to lose his voice in order to inhabit this public space. Whether this is simply a byproduct of the expression of knowledge as power, or is an assertion of power over the people, these library buildings communicate powerful messages.

Fast-forward to the 1960s and we find an entirely different message being broadcast by the branch libraries, sometimes referred to as "bunker" libraries. These buildings are characterized by an aesthetic that expresses economic austerity. Typically built of unadorned, running bond brick or concrete masonry, these one-story buildings present unfriendly, blank facades. Where windows exist, they are often of vandal resistant glass block, or are clerestory strips small enough to prohibit breaking and entering. Not only are they decidedly unwelcoming, but in some libraries, like Baisley Park Library in Queens, the entrance seems purposely hidden behind a brick alcove, furthering the impression of the library as an impenetrable fortress. The interiors are not much better, with low ceilings and painted concrete block walls. What little daylight is admitted through windows is overcome by row after regimented row of fluorescent lights. These depressing buildings are a condemnation of "dangerous" neighborhoods and an expression of devalued citizenry.

Figure 2.1. A "bunker style" library.

Branch libraries built in New York in this first decade of the 21st century stand in stark contrast to the bunker buildings. Guided by the desire for transparency and openness, these buildings seek to engage with dynamic, daylight interiors visible to the street. Facades and massing are meant to both welcome their users, and communicate the library as a beacon for learning. Built at grade, with open plans, high ceilings, and no imposing central circulation desks, these libraries mix the civic pride and aspirational goals of the Carnegies with the current ideal of equity and equal access. At the Glen Oaks branch library in Queens, a broad, welcoming entrance meets the users on the ground plane, next to an open, planted reading area. A natural skylight beacon, modeled after a sundial, shifts across the façade as the sun moves throughout the day; encouraging the library's patrons with the word "SEARCH".

The powerful civic architecture of the late 19th and early 20th centuries, as expressed in the Carnegie libraries, has become a sort of ideal American civic architecture, with its soaring spaces, quality materials, and elevated craftsmanship. However, the forms no longer reflect our societal structures and instead reinforce political and cultural values that are in some ways better left behind. Likewise, the bunker buildings of the 1960s and 1970s are similarly out of synch with today's values. More importantly, they also clearly demonstrate how deleterious a message can be; a concrete example of hostility that lends itself to a disenfranchised populace. The libraries built in the last decade in NYC show how we can learn from the past. They reinterpret strengths and avoid past missteps, while creating a purposeful and aspirational message of openness, equity, and valued citizenry.

Uncovering, Assimilating, and Expressing Meaning Through Buildings

Architects are trained in the practice of design methodology. It is through this practice that they are able to uncover both physical and aspirational program requirements, to tease apart the messages conveyed by previous buildings, and to recombine these elements into a new narrative. In examining the design process for the FDNY Rescue 2 station we can see how this methodology is used to craft the message of a public building.

New York City is proud of its firefighters, and many of the City's fire stations communicate this pride. Like the Carnegie branch libraries, many of the fire stations have a strong turn of the century iconography that has embedded itself in the City's psyche. When the architecture firm Studio Gang was selected through the City's Design Excellence Program to design the new Rescue Two station in Brooklyn, they recognized that they would need to understand the history, culture, and function of the firehouse in order to create a building that captured the civic pride and required functionalism of the traditional firehouse.

Studio Gang conducted a study of over 50 NYC firehouse facades in order to analyze how the pride people have in firefighters is lodged in the elements of the station facades. They found that stations are typically built of solid, dependable masonry with friendly, residential scale upper story windows. The ground level house watch windows indicate eyes on the street, and the traditional red station doors signal the building's function. Straightforward identification proudly states the building's purpose, and every station flies the American flag. That these exterior elements communicate civic pride is imperative for a building that is rarely open to the public, due to the nature of the services it performs. It is up to the facades to testify to the willingness and dedication of the FDNY to serving the neighborhood.

Similar to the way they teased out meaning embedded in fire station facades, Studio Gang also studied the functions of the squad. A great deal of the fire company's time is spent on training. In this case the unit who will be housed in the station is a specialized rescue unit; their training focuses on rescues of both civilians and other firefighters. They respond to highway emergencies, tunnel collapses, scuba rescues, and other catastrophic events. They train to gain familiarity with the specialized tools required by their trade, specific rescue maneuvers, and with the conditions of the city, and of its built environment.

Studio Gang studied precedent training structures, culling key elements that could be incorporated into the building. They also familiarized themselves with the specialized tools required for rescue operations. This research inspired them to create the building as a direct expression of preparedness. Like tools and their precise place on the firetruck, spaces of the building are precisely organized to aid in rapid deployment. The building itself can function as a training tool. The light shaft at the center of the building becomes a sort of inside out training tower; the basement can be filled with smoke for confined space exercises.

Through design methodology Studio Gang was able to take apart precedents and reconfigure them to create a building that recognizes both the role of the fire station within the larger city, and the core mission of the squad. This allowed them to repackage cultural meaning and required functionality into a communicative building, one that does not rely on limits of historic expression in order to send its message. It projects security, safety, and civic pride to the community, and also respect for the importance of the firefighters' livelihood. It speaks of the commitment the city government has to both the neighborhood and to its workers.

Government Support of Good Public Architecture

The messaging of the built environment has a powerful effect on how a person experiences their surroundings. While the experience may be inchoate, the effort put into the messaging must be precise and purposeful. Architects are able to uncover and create messages through the use of design methodology, as shown above. However, they are not necessarily adroit at navigating complex and often archaic government bureaucracies; indeed, few outside of government services are. For this reason, the architect's endeavors must be supported by those within city government who are directly involved in capital improvement. In New York City there are two broad government categories that are set up to support good public design: the management agencies tasked with leading capital improvement projects, like the Department of Design and Construction, and the oversight agencies whose responsibility it is to review and approve of public design projects, like the Public Design Commission.

The Department of Design and Construction

The process of designing and constructing public buildings in New York City is complex. Some city agencies oversee their own capital improvement projects, while others depend on the services of agencies like the Department of Design and Construction (DDC), which is the City's primary design and construction management agency. The DDC provides design and construction management services to many of the city's other agencies, including the library systems, the health and human services agencies, and the uniformed services. In addition to fire and police stations, senior centers, homeless and recreational facilities, and libraries, the DDC builds plazas, streets, and other city infrastructure. With a portfolio of approximately $15 billion, the projects it oversees have a real impact on the environment of the city. (http://www1.nyc.gov/site/ddc/about/about-ddc.page)

DDC is responsible for selecting architects and engineers, creating the project scope based on client agency needs and available budget, and in general overseeing the design and construction process. However, it must cede portions of the process to other city agencies, such as the Office of Management and Budget and the Mayor's Office of Contract Services. These entities do not have design quality as central to their core responsibilities, yet have decision making power over projects.

For example, the Office of Management and Budget (OMB) has control over how funding is allocated within a project, and is tasked with assuring there is no misuse of government funds. They will often rule on specific items within a project scope with little consideration for how that item may affect the project as a whole.

Based on a rule stipulating that the city cannot replace capital work done within the last ten years, OMB might insist that ventilation ducts installed in a building slated for demolition be re-used in the new replacement building despite the fact that it will have a completely different layout. Or they might find painting the interior of a building capitally ineligible because it is defined as a maintenance measure, despite the fact that the project contains a number of new walls. This is not to say that these agencies are not following required rules and regulations, many of which are aimed at preventing corruption and the misuse of government funds. However, these requirements illustrate how the public process is fraught with bumps and bottlenecks that can cause stakeholders to loose focus on design quality.

The Design Excellence Program

In 2004 the DDC started an innovative program called the Design Excellence Initiative in order to address these and other shortcomings. Based on a federal program of the same name, it aimed to improve the design quality of public buildings in the city through attracting high caliber architects and by addressing some of the larger government obstacles to good public design. The initiative advocated an entirely new way for the city to select the architects of public buildings. It also developed strategies to address obstacles throughout the process of producing successful public buildings and spaces. Introduced as a temporary initiative, the pilot proved so successful that it was adopted as a permanent program to be used as the default procurement method for capital improvement design consultant services for the city.

Prior to the Design Excellence Program, the City's procurement laws required that an architect selected for a city project be the lowest responsive bidder for that project. Although fiscally responsible, it is not difficult to see that a procurement method based on the lowest cost services will likely yield low quality services. In addition to the low bid requirement, the standard procurement process required substantial and narrowly defined previous experience, making it virtually impossible for new talent to enter the field. Competition was open only to a narrow subset architects who had previously designed the same type of building. New architects could not win contracts because they had not previously won any contracts. This is another example of one agency's regulations not incorporating the value of another agency's goals. The Mayor's Office of Contracts, in their well-founded aim at preventing corruption and fiscal irresponsibility, did not take into account that new talent and a larger field of competition could raise the bar and simultaneously support the city's architectural community. A more pervasive problem was the assumption that only an architect who had previously built this type of project was capable of building it again.

The Design Excellence program successfully argued for selecting architects through quality based selection. Rather than requesting that the Mayor's Office of Contracts change their policy, DDC avoided the lowest bidder requirement entirely by incorporating what MOCS deemed to be fair and competitive fees into the contract. Architects who responded to the Request for Qualifications implicitly accepted this standardized fee, so that in essence all the bidders were proposing to work for the same fee. Over time, the program's results convinced key administrators that the problem solving capabilities of a good architect were broader than a singular building type, and in fact were strengthened by a variety of past professional experiences.

The Design Excellence program developed other strategies to address bureaucratic obstacles within the capital improvement process as well. For instance, specific goal setting strategies were developed to address the fact that public projects involve a large number of stakeholders, often with competing project goals. Professional peer reviews were introduced both to create better bonds with the City's architecture community and to elevate and broadcast best practices within the City's agencies. Perhaps one of the most effective strategies was creating the new role of the Design Liaison. DDC understood that the Design Excellence initiative would only be as successful as its implementation. The Design Liaisons' responsibility was to protect the design process from bureaucratic pitfalls, and advocate the importance of good public design to stakeholders; to lead architects through the straits of the public process, and to ensure that the many decisions made throughout the process were not made in a design vacuum.

In order to do this, the Design Liaisons were given the authority to operate outside the strict hierarchy of City agencies, navigating both within and outside DDC to remove obstacles and create pathways through the complex process. The sheer number of agencies involved, the range of professional backgrounds represented, and the myriad of often competing priorities brought to each project required that the Design Liaison consistently build consensus and bridge differences.

The Design Excellence initiative also included an annual lecture series aimed at increasing agency staff exposure to best practices in public design, and forging bonds with the City's architecture community. Offering professional continuing education credits and scheduled during normal work hours, the series allowed DDC agency staff to absorb these design excellence strategies, and engendered a seismic cultural shift towards acknowledging the value of design within the agency.

Now in its twelfth year, the Design Excellence Program is considered a success. The new Glen Oaks library, the FDNY Rescue 2 station, and hundreds of other DDC projects produced through this program have created a new generation of public buildings across all five boroughs which convey messages of valued citizenry and have raised the quality of life for countless city inhabitants.

The Public Design Commission

Good design is as much about process as it is about product. Studio Gang used design methodology to create a modern rescue station rich and resonant with cultural meaning; DDC uses it to navigate, problem solve, and build consensus throughout a complicated public design and construction process. The NYC Public Design Commission (PDC) uses it to broker consensus at high levels across the city, to hold agencies responsible to good design, and to guide and safeguard the messages embedded in civic spaces.

Established by the 1898 City Charter, the Public Design Commission's mission is to elevate the design quality of public projects. Sprung from the City Beautiful movement, and originally named the Municipal Art Commission, the Design Commission reviews buildings, open spaces, and art proposed on city owned property. Public buildings, parks, plazas, streets, and bridges all fall under its purview. (http://www1.nyc.gov/site/designcommission/index.page).

The City Charter sets the requirements for the eleven member Commission, and specifically requires that the board consist of an architect, a landscape architect, a painter, and a sculptor; in addition to three lay members and representatives from three of the city's earliest cultural organizations: the Metropolitan Museum of Art, the New York Public Library, and the Brooklyn Museum. Commissioners are typically highly regarded in their field and have demonstrated a commitment to public service and an abiding interest in the welfare of the city.

The Commission has jurisdiction over any capital project on city owned property that has exterior scope, and all public art whether exterior or interior. Any agency that is engaged in public capital improvement in the City is likely to submit a number of projects to the Design Commission each year. The agencies that submit the most frequently include the Department of Parks and Recreation, the Department of Transportation, the Department of Cultural Affairs, the Department of Environmental Protection, the Economic Development Corporation, and the DDC.

In essence, the Commission is a rigorous and binding peer review, and demands that projects be grounded in the tenets of good design. Its focus is to make sure the design choices have been made intentionally and in a well-informed manner, so that they express both the stated goals of the project, and those of the City as a whole. PDC decisions are binding—no project on City owned property can obtain a construction permit from the Department of Buildings without a certificate of approval or waiver from the PDC.

As with any oversight agency, pushback against PDC does occur, and ranges from a baseline dread of yet another step in an already complicated process, to the outright belief that design either doesn't matter, or is wholly subjective. Particularly in agencies that do not have a robust Design Excellence program, quality design can fall victim to the many complications outlined earlier in this chapter. In these cases, the PDC can act in place of the Design Excellence program. However, since it only reviews designs once they have been developed, the PDC can only act in a reactive manner. This can often cause pushback from the agencies proposing the designs, especially those who have not given the proper priorities to creating a good design in the first place, and who now are concerned with backtracking in order to right this oversight. When this occurs, PDC requests that the project be resubmitted with revisions that address their concerns. Often the PDC's commentary is broad enough that a resolution can be quickly reached that responds both to PDC concerns about the quality or appropriateness of the proposed design and the concerns that are causing the resistance.

The Commission also comes under fire from parties with priorities that are seen to be in conflict with its mission to elevate the quality of public art and design. The primary underlying dissatisfaction is too often a misperception that good design comes at the price of schedule or budget and is of less value than these other determinants. That these detractors are not able to see the positive impact that the built environment has on the wellbeing of the city underlines the necessity for design review Commissions like the PDC.

While disagreements do occur over design issues, more often as not projects falter due to issues relating to coordination across agencies, a lack of consensus among all involved stakeholders, or political or even public dissatisfaction with a project.

PDC's approval is a regulatory means by which to keep agencies to their word when other forces prevail on a project's design. PDC review is also often the only point during the design process that requires all involved city agencies to coordinate around a project. Written approval from the Commissioners of all involved agencies is required as part of a project submission. This explicitly ensures that agencies have coordinated their priorities and are all satisfied with how the design addresses them. If the organizations and agencies do not demonstrate a coordination of goals, a centralized vision, and clear communication, the PDC has the authority to reject their submission as incomplete.

Due to the way funding is segregated, as well as the rules governing consultant selection, there are sometimes separate projects being carried out concurrently on the same building, sometimes even lead by different agencies. In one case, two separate submissions for a small library were submitted to PDC at the same time, one for a roof replacement and one for the installation of ventilation equipment on the roof. The roof installation was slated to occur first, with the equipment installation to follow a few months later. The manner in which the equipment was to be installed on the roof would have voided the roofing warrantee; because the PDC demanded coordination between these projects and noted this error, the agencies were able to correct it before time and cost delays were incurred. The PDC review process can ensure good design methodology and coordination across city agencies through the submission materials it requires. The benefit of this cannot be overemphasized. In a city as large and complex as New York, miscommunication and lack of coordination can often dog a project well into construction. It is a well-known that problems solved during design cost far less money and time than those left to linger into construction.

PDC is generally protected from the changing priorities of individual administrations because its jurisdiction is part of the City Charter. However, the degree to which the Commission is empowered to act on its authority varies in relation to an administration's overarching interests. Public opinion plays a role as well; currently, growing public awareness of the power of good design on cities has bolstered PDC's authority. In order to capitalize on this, the Commission must use its power responsibly. This means sticking to the known tenets of good design, which can be objectively criticized, and respecting the limitations of schedule and budget without allowing them to overshadow the importance of design.

The same goals and values we see in the DDC's Design Excellence program exist at PDC, in a broader range and at a higher level of city government. They also involve a larger slice of the city, including political representatives, community organizations, and the general public. Community Board review is required before Design Commission will calendar a project for review, and the PDC hearings are open to the general public. These public hearings are often the sole point during design when public testimony can be made; the general public, community members, advocacy organizations, City Council members and Borough Presidents can and have all testified on specific projects. This testimony can uncover all sorts of conflicts, from disagreements within communities on project goals, to objections over proposed uses of open public space, to concerns that the city has not adequately considered specific histories or elements of cultural relevance. While testimony often involves areas outside of the PDC's focus on quality of design, the public hearings are nonetheless an opportunity to bring some transparency to the public capital process, and to bring the importance of quality public design to the attention of the greater public.

One of the most valuable characteristics of the Design Commission is its unique vantage point; it has an eagle eye view of the projects going up on city owned land across all five boroughs, with a hundred year old archive of projects on 7,000 city owned sites. Far from gathering dust, the archive is actively used during the review process since every project submitted is subject to an archival search. In this way, the Commission is able to bring a depth and breadth of knowledge to bear on specific projects, and is also in a position to understand the messages of the whole city, from streets to parks to libraries to fire and police stations. It is uniquely situated to make sure the city is intentional with its message. The Public Design Commission is an unusual instance where government regulation aids artistic expression. As the Landmarks Preservation Commission is tasked with protecting the historic culture of the city, the Public Design Commission is responsible for guiding the messages our public buildings will tell in the future.

The Design Commission review process is imperative for successful delivery of well-designed projects. However, it cannot create good public design on its own. Projects are typically reviewed only twice during the process; once at a schematic level, and again at final design. The Commission only comments on designs once they are conceived, unlike the Design Excellence Program's proactive guidance. Furthermore, PDC does not have purview over land use decisions, zoning requirements, or programming, which are determined by the Department of City Planning, the Department of Buildings, and the client agencies. The symbiotic relationship between the DDC's Design Excellence program and the Public Design Commission review has already had a profound effect on both the process and product of New York's public work. Further coordination with other agencies would undoubtedly make the city's support for excellent civic buildings even stronger.

A larger obstacle to Design Commission, the Design Excellence program, and any other program in support of good city design is lack of awareness of the significant positive impact design has on cities. While awareness is growing that good design has value, it is not necessarily clear that there are objective characteristics of good design, and many are still unaware that these can greatly impact the built environment they live in. Add to this the many real and competing priorities in public capital improvement and it becomes clear why people within city government must take an active role in good design if they want their cities to communicate a healthy message that reflects their values and helps to create an equitable, socially resilient, and culturally meaningful public realm.

Events like the Public Design Commission's annual Design Awards bring attention to the public buildings that best demonstrate these attributes, as do collaborative events with the the City's design institutions, such as a recent series of panel discussions focusing on public design and equity that PDC developed with the Architect's Institute of America's New York chapter. Design Excellence can serve to educate the many people involved in the public capital process who are not aware of the powerful benefits good design brings to a city and to promote understanding of how good civic design promotes physical and mental health, and overall quality of life.

Conclusion

Public buildings and spaces express the degree to which city government values its inhabitants, as well as the level of investment the government is willing to make in its people. Buildings can carry messages of welcome or they can turn their backs on inhabitants. They can and should convey a sense of civic pride, encourage people to aspire, and impart a sense that those who inhabit them have worth. Since these messages are powerful, and affect how people feel and behave, the city has the responsibility to craft them carefully and purposefully, so that they communicate values such as openness, recognition, and community.

Public capital improvement is by nature complicated, and involves a large number of stakeholders. Champions of good design would do well to take these public sector challenges seriously, understand what is at their core, and advocate for the governmental tools necessary to combat these obstacles. The Public Design Commission, the Department of Design and Construction, and the Design Excellence Program champion good civic design from within government. They maintain focus on questions of design amidst the clamor of competing concerns and forces. Furthermore, they can be used to educate those in positions of power that design is a potent tool in the fight for equity, social justice, and economic health.

NYC is unique, both in the size and complexity of its government structure, and the amount of creative capital is can draw on. However, many of these tools can and should be adapted to other cities, since the obstacles to good public design outlined in this chapter are in evidence across the United States. If we cannot accept and act on the fact that good public design creates cities that thrive, we will not build successful cities. As New York has shown, our city governments can and should be powerful players in creating more dynamic, responsive, equitable, and healthy places to live.

CHAPTER THREE

The Roles and Regulation of Urban Graffiti as Communication, Art or Criminality

GENE BURD
Belo Center for New Media, Austin, Texas

The ghost writers of graffiti trespass city borders and boundaries as they practice an outlawed non-commissioned "art" outside of the rules and regulations of art galleries and beyond review and control by benefactors, editors, publishers, politicians and police in contested urban spaces. They "speak" in the streets and pathways with their messages beyond the juries and judges as they seek to connect to the urban community without face-to-face confrontation in the geographical space and place on the urban commons. Their messages are fixed on urban structures and sites or hitchhike on vehicles in fluid mediation, evading immediate feedback and challenge by viewers. Fundamental and perplexing issues are raised by graffiti: Should it be prevented, controlled, or sanctioned? Is graffiti protected expression or prosecutable crime? (Young, 2013).

"Art crime" is one of many complex legal issues emerging from graffiti. From a regulatory standpoint, graffiti-related rights, rules and regulations vary greatly from country to country and city to city. Frequently considered criminal behavior, the seriousness of the graffiti offense ranges from petty crimes/misdemeanors to felonies carrying significant jail sentences. But graffiti is more than a potential crime, it is also potentially valued art. The artistic value of graffiti has been recognized by some evidenced by the practice of inviting graffiti artists into some cities as a tool of revitalization. The work of graffiti artists has been welcomed by art galleries and museums. The works of Haring and Banksy sell for millions of dollars.

Legal battles continue to be fought over the boundaries between the right to an artist's free expression and a property owner's right to use and enjoyment of their own property. In addition, graffiti art has been the subject of copyright litigation.

Much of the world of graffiti grows and thrives "outside the lines" of the established order amidst anonymous, angry adolescents who are defensive, guarded, resentful and rebellious "outsiders" whose "drama of deviance" fits modern urban life (Becker, 1963). Their habitat is one of risk, danger, and exploration of status of "us" and "them" in a battle for boundaries on the urban turf. Small wonder that they alarm the authorities. Graffiti was and remains a way for youth to trespass restrictive borders, from pastoral and rural tree carvings outside the caves and tents to initialed concrete and hop-scotch drawings on city sidewalks. Youths seek acceptance in space, invading it with threatening, forbidden and often obscene taboos in bathrooms and on public walls in quick, short and irresponsible grammar and then silently retreat into safer enclaves protected by identity with race, sex, class, status, or ideology. Even Pompeii's legitimate "youth club" had its walls "to keep the inside place safe, and the enemy outside baffled" (Lindsay, 1960, p. 87).

Not limited by the restrictive rules of parental or urban authority, child-like communication and its adolescent "art" imposes themselves upon what it sees as restrictive structures which "wall" them out. Their words, format and colors are often brief, exaggerated, unambiguous and undisputable in child-like frustration. "Such graffiti are announcements of one's identity, a kind of testimonial to one's existence in a world of anonymity" (Abel & Buckley, 1977, p. 16). Those who write it usually take no responsibility for their actions and are free to attack and yet remain invisible and anonymous except perhaps outside their own group (Klinghoffer, 1993). With the child, "Everything in life is either forbidden or compulsory" (Nilsen, 1980, p. 239); and graffiti is "a form of communication that is both personal and free of the everyday social restraints" (Abel & Buckley, 1977, p. 3). "Graffiti writers are not afraid to take liberties with the forms of words". They play with puns, alter existing expressions, use planned or spontaneous dialogue, invert expected word order and "get maximum communication value with a minimum number of words" (Nilsen, 1980, p. 236).

In response, graffiti control efforts have been massive. These include arrests, fines, hot-lines, live cameras and graffiti "busters" and guards to report and catch graffiti writers; paint-outs, sandblasting, taxes on spray cans; special cleaners and removers; resistant and protective wall paints, and even counselors for offenders. Their success has been limited, in part because it is difficult to prevent the movement of words across physical boundaries in the city. The subway commuter who sees graffiti feels "assaulted continuously … by the inescapable knowledge that the environment he must endure for an hour or more a day is uncontrolled and

Figure 3.1. Graffiti artist at work.

uncontrollable, and that anyone can invade it to do whatever damage and mischief the mind suggests" (Glazer, 1979, p. 4). Although as a crime it may be "relatively trivial", despite the "ever-present sense of fear … the *sense* of a dangerous place is different from the *reality* of a dangerous place" (Emphasis in original). Graffiti "contributes to a prevailing sense of the incapacity of government, the uncontrollability of youthful criminal behavior and a resultant uneasiness and fear" (Glazer, 1979, pp. 5, 10–11).

Art of Crime

From pre-historical to post-modern times, graffiti has been used to mark contested space, with participants announcing their independence from authority so as to control new forms of expression and "own" a part of the landscape. Such territorial wars have defined community turf through competition to control visual and print symbols that cross or trespass walls separating "us" and "them", "insiders" from "outsiders"—from walls of caves to the Berlin Wall, from the pillars at Pompeii to the Tower of London, from buildings in Manhattan to barrios in Los Angeles, or barricades in Moscow. Graffiti writing or drawings scribbled, sprayed or scratched on walls or other surfaces in public places is generally considered a nuisance flourishing and perpetuated by homeless truants in railroad box cars in the 1920s and by street gangs and hip-hop artists in the 1960s into the mid-1970s with police arrests along with legislation penalizing graffiti writers. Big U.S. city "wars" on graffiti in the 1980s brought prosecutions for vagrancy and vandalism on properties, fines and jails for graffiti writers and bans on the sale of spray paints (*Wikipedia*, 2015; Davies, 2012).

By the 1990s and into the next century, scholars and artists began to question the notion of graffiti as vandalism or crime and suggested that "street art" was a "variant" of graffiti but "the law does not distinguish between writing and art and so graffiti writers and street artists are equally subject to prosecution" (Young, 2013, p. 100). That "socio-legal response" is complicated by the "enigmatic nature and insular culture" of the art form of graffiti calligraphy which is "hard to read and most people never see a graffiti writer at work" since the writing usually appears overnight, and its author remains unseen (Young, 2013, p. 100). Contested issues abound around the ownership, propriety and authority of un-commissioned words and images often scrawled and scribbled in deteriorated places often symbols of disorder and incivility. Also, there are the difficulties in proving physical damage and negative consequences from graffiti beyond symbolic and hysterical moral panic; plus what is public and what is private and who owns or has authorship to the city. Also of interest is how many graffiti artists "regard illegality as giving meaning to what they do" plus "the ways property and authority in urban space are differently imagined by law and by the street artist" (Young, 2013, pp. 107, 123). An outlaw urban communication culture of outcasts marks graffiti boundaries and confronts political and legal authorities defining and dealing with a legitimate revolt and resistance against what are seen as walls of social, political and architectural elitism marking urban territory and the public domain (Brighenti, 2010). Changes in urbanization and technology have also cultivated and accelerated a "guerilla aspect" to the graffiti artists as they are "involved in acts of reclamation" to construct "discourses of resistance" thereby "reclaiming the 'space' which they feel belongs to them and which is vital to their sense of both freedom and identity" (Bennett, Kendall, & McDougall, 2011, p. 31).

Efforts to regulate this "urban communication" as a crime has been extensively documented in over a decade of studies. New York City played an historic role in the rise of modern graffiti and has also been a leader in regulatory initiatives. On the state level, New York Penal Law §145.00 clearly criminalizes graffiti stating "a person is guilty of criminal mischief in the fourth degree when he or she, having no right to do so, intentionally damages property of another person, intentionally participates in the destruction of an abandoned building or recklessly damages property of another person in an amount exceeding $250." §145.05 of the State Penal Law states that a person is guilty of criminal mischief in the third degree when with the intent to damage property of another person, and having no right to do so, he or she damages property of another person in an amount exceeding $250. §145.10 states that a person is guilty of criminal mischief in the second degree when, with intent to damage property of another person, and having no right to do so, he or she damages property of another person in an amount exceeding $1,500. In New York, vandalism is covered under the legal definition of "criminal mischief." Graffiti is considered a criminal mischief of the third degree crime in New York and several other states throughout the U.S. In New York this is a Class E felony, which is punishable by up to 4 years in state prison and fines from $250 fine to $1,500. Under New York Penal Law Criminal Mischief in the Fourth Degree is an "A" misdemeanor punishable by up to one year jail under Penal Law 145.60.

Further, Possession of Graffiti Instruments falls under New York Penal Law 145.65 which is s a "B" misdemeanor punishable by up to ninety days in jail.

In New York City, §10–117 addresses "Defacement of property, possession, sale and display of aerosol spray paint cans, [and] broad tipped markers and etching acid prohibited in certain instances." This law prohibits writing, drawing, painting or making any mark on a public or private building or other structure or any other real or personal property owned without consent. Further it criminalizes possession or sale of the instruments used for graffiti. Under §10–117 (b): "No person shall carry an aerosol spray paint can ... with intent to violate the current (graffiti) laws in place". This is a New York City law, where people can be fined up to $500 just for the possession of a spray paint can if a cop assumes that you will/ have violated a law. Under §10–117 (c): "No person shall sell or offer to sell an aerosol spray paint can, [or] broad tipped indelible marker or etching acid to any person under eighteen years of age." Store owners in New York are not allowed to put cans of spray paint on display, either—they can only display facsimiles of the cans. In New York City Title 10 §117.1 creates an Anti-Graffiti Task Force. Title 10 §117.2 of the New York City Administrative Law Code allows the Mayor to offer and pay a reward, not exceeding $500, to any person who provides information leading to the apprehension, persecution or conviction of any person who vandalizes property.

Graffiti artists can also be charged with criminal trespass. For example, under the Utah Criminal Code Title 76-6-206, Criminal trespass specifically includes anyone entering on property who "(i) intends to cause annoyance or injury to any person or damage to any property, including the use of graffiti as defined in Section 76-6-107."

Graffiti is considered a form of vandalism in many places around the world. For example, in Western Australia, under the Criminal Code Act Compilation Act 1913 amended in 2010 not only addresses graffiti as vandalism but includes the banning of the sale and supply of spray paint (State Graffiti Task Force, 2016). https://www.goodbyegraffiti.wa.gov.au/Police/Graffiti-and-the-Law.

In England and Wales graffiti falls under the Criminal Damage Act of 1971. Under Section 6 graffiti is a criminal act that can be fined up to £5,000 if the damage they have caused is less than £5,000. Damage in excess of £5,000 is referred to the Crown Court which can impose greater fines. Community service can be the penalty for youthful offenders. Some laws link graffiti and gang related activities and further associate this with criminal behavior of minors (e.g. California Government Code §38772. Defacement of property by graffiti by minor).

In some municipalities penalties are distinguished based on the amount of property damage caused. In some situations distinctions are made between graffiti on public property vs. private property. Prior convictions can also determine criminal penalties. Graffiti committed by a juvenile vs. an adult can also determine the charges brought and penalties imposed upon conviction since generally juvenile courts have much broader discretion than adult courts. For example, if you are convicted of a graffiti crime in the state of Nevada, the penalty ranges from as little as fine of between $400 and $1,000 and serving 100 hours of community service, to as much as five years in a state prison and a $10,000 fine (NRS 206) https://www.leg.state.nv.us/nrs/NRS-206.html.

However, the argument is frequently made that graffiti is a victimless crime. Yet the response is often simply the assertion of the rights of property owners. Some cities have sought ways to balance graffiti and street art management. In Melbourne the city sought to a policy focused upon notions of "negotiated consent" to the presence of graffiti or street art and "zones of tolerance" with varying degrees of self-regulation by graffiti writers or external regulation by police and council authorities (Young, 2010)3.

Elsewhere governments have gone so far as decriminalizing street art. The Brazilian government passed federal law 706/07 which removed criminal penalties for defacing urban buildings and monuments if done with the consent of owners. As a result, the mayor of Rio legalized graffiti on city property. Graffiti initiatives have been embraced by Rio as a form of city beautification, deterrent to crime and "career opportunity for youth in low-income neighborhoods" (Olivero, 2014).

Beyond being an issue of criminal law, graffiti has also raised questions in the realm of copyright law. There has been some debate as to whether illegally-created graffiti is entitled to copyright protection. Graffiti satisfies the requirements for copyright protection as an original work set in fixed tangible form. Villa v. Pearson Education, Inc. is the case on point that that directly addresses the copyrightability of graffiti (Villa v. Pearson Education, Inc, 2003). 2003 WL 22922178 (N. D. Ill. Dec. 9, 2003). There the court did not state explicitly that a finding of illegality would preclude copyright protection (Davies, 2012). In the *Maine Law Review* article "Art Crimes?: Theoretical Perspectives On Copyright Protection For Illegally-Created Graffiti Art" Jamison Davies asserts "Graffiti art has a good, but as yet untested, claim to copyright protection" (Davies, 2012, p. 38). While he concludes that copyright is not all that important in the realm of providing incentives to this form of creativity, the growing recognition of this art form evidenced by gallery and museum exhibitions suggest an emerging legal realm for graffiti.

The International Municipal Lawyers Association's (IMLA) developed a comprehensive Model Anti-Graffiti Ordinance. Provisions were based on ordinances in Los Angeles and San Diego, California; Wilmington, Delaware; Chicago, Illinois; Boston, Massachusetts; Portland, Oregon among other cities as well. Cities including Indianapolis, Indiana have adopted portions of the IMLA Model Anti-Graffiti Ordinance. The purpose of this ordinance is to prevent the spread of graffiti vandalism and establish a program for removal (International Municipal Lawyers Association). (http://www.anti-graffiti.org/images/IMLA-Grafitti-Model-Ordinance.pdf). It positions graffiti as a criminal nuisance, a misdemeanor, and provides for removal by either the perpetrator or by the property owner or city. The IMLA Model also includes a provision for the establishment of a trust fund associated with the cost of removal. The regulatory approaches surveyed undervalue the powerful communicative functions of graffiti as expression.

Graffiti as Urban Expression

The rules and regulations of communication in urban space are crucial for the collective graffiti experience in cities, the geographies of who has rights to the city, and who owns private and shared urban space in the battle for turf and territory in the flow or communication in the city. Framed as nuisance, vandalism and as transgression rather than legitimate public communication graffiti has been described as an "intrusive, emblematic, opportunistic", premeditated and purposefully "out of place" mode of "outlaw communication" in an "unruly alternative public sphere" of "forbidden inscription". It has been seen as a "symptom of the failure of the limits of permissible or legitimate communication within the public sphere", which "challenges us to think about the relationship between people and places" which is "crucially implicated in an emergent spatial politics of entitlement and belonging within the city" (Back, Keith, & Solomos, 1999, pp. 69, 98–99).

The "analysis of graffiti is a valuable opportunity for communication researchers ... to view graffiti as valuable ... unobtrusive data that reveal conflicts and tensions that dominant groups endeavor to suppress" and which "augments other genres of communication (e.g. language, metaphors, ceremonies and stories) that focus on the process notions of representation, oppression and resistance" (Rodriguez & Clair, 1999, p. 14.).

Graffiti, like art, is "a profound iterative form of urban expression" which "has endured as the most archaic form of written communication" and "is not merely the sign or the effect of an underlying ethological community; rather, it is the embodied social landscape. It draws attention to how communities are manifest and articulated in a necessarily incomplete, uneven, and contingent fashion" as "we need to be able to read the writing on the wall, for in these opaque and arcane inscriptions, we can find both the limits of public discourse and the outlaw voices that threaten to be heard" (Back et al., 1999, pp. 97–99).

Graffiti is an excellent example of urban communication outside the boundary lines of the restrictive forms and content of city life. Like a child "outside" seeking to get "inside", graffiti trespasses and crosses over to challenge and violate the authority rules of adults. It is also an ideal unobtrusive method (Raymond, 1989; Rhyne & Ullmann, 1972; Stocks, Dutcher, Hargrove, & Cook, 1972) to study conflict in cities and communication and possible permanent institutionalized post-modern resistance to the architectural landscape. It illustrates the use of space, which "speaks" and "communicates" territoriality in a "silent language" (Hall, 1959). The term "graffiti" has long been "seeking acceptance into the language" in the urban glossary (Abrams, 1971). Its use of space is like a "hidden dimension" of communication embedded in the personal and social perception of physical reality and the psychological need for boundaries. It is a "language" as easily "read" as if printed and spoken. Nevertheless, "we treat space somewhat as we treat sex. It is there but we don't talk about it" (Hall, 1969, pp. 6, 188, 194).

If such place-makers are "relevant to the people who are going to experience them", they are not "monumental arrogance" that invites "protest with wanton acts of vandalism", but provide "a declaration of public value" (Fleming & von Tscharner, 1981, pp. 17, 112). Understandably, graffiti by "taggers" and gangs becomes competitive "territorial markers" which define distinct turf and boundaries and claim space for the powerless and dispossessed through "an open declaration on the walls" with messages that are "internally supportive and externally aggressive" (Ley & Cybriwski, 1974, pp. 501, 504).

As one writer put it "Today's graffiti are tomorrow's headlines". The grammar is short and to the point, sometimes unqualified, and combines anti-social thought with disfigurement of someone else's property thereby creating an "emotional orgasm" for the writer (Abel & Buckley, 1977, p. 74).

In respect to boundaries, the suggested relationship of graffiti to arrested adolescent sexual and mental development is important. It is said that graffiti represent thwarted human interests (Opler); and toilet graffiti in particular is evidence of aggressive and hostile impulses due to sexual repression (Solomos & Yager, 1975). It's a place for males' coarse, phallic bragging, but for females, romantic inscriptions (Wales & Brewer, 1976); a place to smoke (Landy & Steele, 1967); and "one place within public spaces where girls can be 'public' but 'invisible'", yet record sublimated sexual desires and yet be safe from an outside where they are denied their rights (Carrington, 1989, p. 92). For troubled and rebellious mental patients, graffiti helps them communicate with others by both building and tearing down "walls" around them (Bayer; *Science Digest*, 1969, 1974).

Graffiti are "representative of the discordance of normal development needs in an urban environment that does not recognize those needs or encourage the fulfillment of potential. It is the attempt to interact with the available social environment in this unfolding developmental process" and the special names graffiti writers adopt serve to separate them from adults and "celebrate the transition from the narrowly defined role in the family to a more widely experienced world" (Feiner & Klein, 1982, pp. 48–49). In the battles over urban space, place, boundaries and borders, the pathology of the city is often blamed for the distortion and failure of communication. The "urban sink" (Hall, 1969) is blamed on city poor design, ethnic and spatial separation, loss of place, crowding and high density.

In Russia, for example, as the authoritarian Soviet Union lost power, graffiti helped youth express antagonisms and define the social space of their hang-outs in their new urban subculture in back courtyards and abandoned basements in Moscow (Bushnell, 1990). Adolescent graffiti writers there "constitute a sharply demarked linguistic community" (p. 163) and a deviant culture of music and politics. It is a country with a long history of graffiti from 19th century Crimean crags to prisoners of the Czars and the KGB. Russian mothers gave graffiti chalk as toys to pacify children to write on sidewalks and even do school math on walls (Bushnell, 1990, p. 163). Angry and youthful Hungarian graffiti writers, restricted by speech censorship, spread their anonymous messages to the general public across "the urban landscape" in violation of political authority and beyond city boundaries into suburbs (Brown, 1995). At Ellis Island, multi-lingual immigrants wrote graffiti as they left an old world and entered a new one (Brooks, 1989).

As "a worldwide network of messages, meanings and symbols … surge across state lines and national boundaries", youths write graffiti in "an expression of the longing to be somebody in a world that is always reminding you that you're not" (Chalfant & Prigoff, 1987, p. 7). It is an attempt "to leave a mark on exotic space, to make a claim to the world outside." as "You go somewhere and you get your name up there and people know you were there and you weren't afraid" (Horn, 1976).Whether markings on a police car, airplane or on the side of an elephant, "The more perilous a locale and the greater the danger of being caught, the greater the recognition given the graffitist by the peer group" (Abel & Buckley, 1977,

p. 140). Subways appeal for their "high visibility, the huge potential audience, and the link with other like-minded kids throughout the city", especially "on the trains, in the dark tunnels, where writers encounter danger, high voltage, cold crushing wheels, giant hurtling monsters" (Chalfant & Prigoff, 1987, p. 9).

Modern cities plan and protectively promise a safe and risk-free urban life, but graffiti shows that architects and city planners may be out of touch with how others are forced to live in what they construct. Some urban critics argue that most environments (i.e. playgrounds, airports, schools) "do not work" in terms of human interaction. There are no "loose parts" but mainly fixed pieces that are "clean, static and impossible to play around with" except for professionals who "have all the fun playing with their own materials" (Nicholson, 1972). As a result, the argument goes, both "children and adults and the community have been grossly cheated" as a majority are not allowed to "experiment with the components of building" a world in order to make experiments. Urban landscape critic, Grady Clay, (1987) goes so far as to suggest that the "all-safe city is the all-dead city … flat, predictable and insurable" but kids "know that if you always play it safe, you'll get no place fast." He thinks "part of the excitement in cities is danger … constructive and healthy dangers" whereby we "learn to measure risk against dare". He sees value in the "wilderness of natural danger", those "crummy places … back-waters away from the rigid regularity of just-so streets and sidewalks and fenced property lines (and) flat asphalt playgrounds" (Clay, 1987, p. 30).

In dangerous forbidden subways, the "powerful machines are tamed by placing one's name on them; the name celebrates victory and possession, like one's brand on a wild steer." This is the period of life in which issues of control are powerful and immediate, particularly with regard to those impulses and feelings within that emerge with great intensity and threaten to become uncontrollable or to conflict with the outside world. Tempting fate and dealing with the powerful projections of inner feelings and sublimation enhance the ability to control these inner powers (Feiner & Klein, 1982, p. 49).

A scholar of urban communication, Richard Sennett, notes the interface of boundaries, borders, youth, urban space and identity:

> The aggressions of graffiti reveal a difficulty faced by all urbanites who are obliged to press themselves upon others in a place without a hierarchy of established recognitions. Who counts, who even exists—these are matters which are established in cities by effort, an effort usually involving struggle, a struggle mobilizing aggression. In a way, what the slum boys did to subways, men in suits do every day in offices, making their impress upon others, the victories lasting only until the next bid for attention, the hustling concentrated in a place in which so many people unknown to one another know they are mutually contending … It is what endless becoming means in crowds; there is a constant struggle for recognition which degenerates into aggression. In a

> city there is no small-scale, rooted village existence which frames who people are; the streets are filled with strangers prepared to resist rather than accept. More-over, the very signs of identity degenerate in our city. (Sennett, 1990, pp. 17–18)

Drawing new urban boundaries for communication in space has been difficult for both youthful graffiti writers and older established authorities who are reluctant to compromise control of urban communications turf. "What is seen by one side as a law-and-order problem requiring responses appropriate to delinquents is seen by others as a generous outpouring of creative energies upon an ugly and demoralizing environment" while to others, graffiti is creative jazz or the abstract expressionism of Jackson Pollack with its emotion and spontaneity, while to others it means a city's symbols are controlled by criminals (Feiner & Klein, 1982).

Norman Mailer defended graffiti as a transcendent folk art saying "There is always art in a criminal act" and graffiti writers live "through the stages of the crime in order to commit an artistic act" and maybe the discomfort felt about it "is the herald of some oncoming apocalypse less and less far away" (Mailer, 1974, pp. 79, 88, 158, 158). One notion is that "a great deal of the finest, keenest, and most satirical and witty writing being done today is not necessarily in books, magazines and newspapers—but on fences and walls" (Mockridge, 1968, p. ix). Graffiti has preceded today's headlines and "The city is like a newspaper anyway so it's natural to see writing all over the place" (*New York_Times Magazine*, 1999, p. 32). Furthermore, many graffiti writers in New York go to Times Square—the advertising billboard and media center already aglow with its visual chaos (Feiner & Klein, 1982), and a place that tolerates adolescence, diversity and deviance.

The question remains whether the mainstream society of art and communication can somehow incorporate graffiti and "can toleration be caught up in the process of annulling difference. What is tolerable can be controlled; the interest of graffiti, as an extreme case, is that they show that toleration can revalue the hostility of protest so that it becomes part of the dominant order's endless debate with itself about its own legitimacy" (Sennett, 1990, p. 22). In this sense, young graffiti writers "have generated for themselves an activity that in its way, is very mainstream America" and it is "helpful to look at the similarities with all adolescents—cooperating and competing with peers" and "finding a way to be excellent, defining boundaries between the worlds of children and adults" (Feiner & Klein, 1982, p. 53). Moving from toilet laureate to public laurels is not so simple or quick. The sexual metaphor of jazz became accepted in music. Urinating a message in the snow is difficult in air polluted cities; and the spray "bomb" is "no longer necessarily illegal or anti-social" but has become a mainstream act (Chalfant & Prigoff, 1987, p. 8).

Efforts to blend these worlds have been encouraged by defining graffiti as legitimate urban communication in various formats: public art through writers' clubs and corners; art dealers, collectors, murals, galleries; billboards and posters on abandoned walls and garbage trucks, public service messages on crime, gang violence, AIDS, drugs on tee-shirts and graffiti bulletin boards (supplied with pencils and stencils); aerosol advertising, and photographed graffiti as art along with artifacts in museums, where segments of Berlin Wall graffiti have sold for $30,000!

In Paris, colored stencils reproducing images of urban life "tattooed" on walls throughout the city have been preserved in photographs (Huber, 1986). Another variation of this interface and overlap of graffiti with legitimate art and advertising is evidenced in the many U.S. graphics collectives which promote, produce, collect and display posters, stencils, banners, stickers, and billboard "corrections". They borrowed from the tradition of ancient satire and lampoons of authority figures; 19th century use of wood fonts, engravings and etchings in cartoons; the "hit and run" graphic "attacks" by anti-Nazi youth; poster slogans used during the New Deal; "graphic guerrilla" techniques of the 1950s; the use of xerox, photo-copying and psychedelic colors in the 1960s; and the new technologies of computers, faxes and direct mail in the 1970s long before the arrival of new interactive Internet technologies a generation later.

These "angry graphics" proposed solutions to social problems through political action on issues like pollution, abortion, AIDS, housing and bad architecture, war, race and sex, crime, police, war, civil rights, health, and the environment. "In many cities there are volunteer snipers who go out into the night with a sheaf of posters and a bucket of wheat paste. These are not vandals. This is not the lunatic fringe. This is evidence that participatory democracy is alive" (Jacobs & Heller, 1992, p. 1). These "snipers" cover, switch and place counter messages at night on billboards and walls, especially plywood and in active grass-roots urban neighborhoods. These appear "surreptitiously, illegally, mysteriously" in "a much blunter, more argumentative form than their daytime counterparts". "Think of it as 3:00 A.M. talk radio made visible", wrote Karrie Jacobs (1992) design critic and editor. This form of public display was called "sniping" and the walls were called "hoardings". They were repeated and some survived and others were torn down or pasted over. These "exhibitions" were "neither judged nor curated. It's a show that's hung and re-hung nightly, mostly by professional posterers, the pirates of the outdoor advertising world", Jacobs wrote.

While space rules apply to legitimate outdoor and transit advertisers, "the city's walls and fences operate beyond the forces of law and propriety … If we think of the fences as a kind of outdoor magazine, the commercial posters are the advertising and the political posters comprise the editorial content. The political posters are missives from people with a message, snipers with a chip on their shoulders." Jacobs explains this mix of political graffiti and what some might call vandalism, as being on "the cusp of art and graphic design". Posters are supposed to shout, but these little posters whispered (Jacobs, 1992, pp. 13–14). Jacobs elaborated further on this media mix:

> As daytime discourse gets thinner and more banal, year by year, as fewer genuine sentiments and opinions get voiced by politicians, and bold political statements vanish from mainstream media, street posters are becoming the one medium in which controversial opinions can find a general audience. … Political posters can be read as an expression of frustration from artists, designers, and regular people trying to transform a political system that seems thoroughly insulated from individual action, a system that seems moribund. The posters that appear on the walls in our city in the night can spark real political debate, maybe night discourse is the last form of political debate. (Jacobs, 1992, p. 14)

Political graffiti has had mixed reaction from graffiti "artists" who sought more permanence than ephemeral slogans (Hougan, 1972). Also, commercial graffiti art galleries failed to become part of the larger art community which had different standards, and lacked "the sociological and psychological conditions which originally motivated the youths … The social structure which the writers had developed could no longer survive outside their own subculture" (Powers, 1996, p. 141). Graffiti art was popular because it gave insight into exciting, dangerous, illegal, and hazardous inner city life on the edge but that subculture was subverted by the dominant culture with its own objectives, rules and history and altered it to conform to mainstream expectations (Powers, 1996, p. 142). A similar finding in a study of the graffiti careers of muralists and taggers found their work gained some recognition and fame among gallery and gang patrons, but their art and deviant careers were still tied to local mentors, peers and audiences (Lachmann, 1988).

Graffiti-like political art and related political wall murals appear to be grounded by geography, history and a mural tradition, as in Latin American and Southwestern U.S. regions. Graffiti as political communication also flourishes when the total society is at war, and a mixture of multiple militant media involve participants at the street level rather than mainly being directed from a single large urban media center. Writings on the Berlin Wall in divided Germany during the Cold War are one example (Rice, 1991; Tillman, 1990). Another is war-torn Northern Ireland, where unionists and nationalists used the popular culture of posters, paintings, post cards, cartoons, book and album covers, songs and poems along with political wall murals (Rolston, 1987). In Beirut, Lebanon's 1975–1982 civil war, militants used a variety of media (posters, rallies, radio, graffiti) to create "one giant mural journal" as a "powerful medium of street communication" (Koleilat, 1985).

In Buenos Aires, Argentina, political graffiti and wall painting is an alternative communication system (Chaffee, 1989). In Mexico, wall graffiti is used as government propaganda to "mobilize the populace to a greater awareness of their social responsibility as Mexican citizens" (Grieb, 1984). Graffiti wall writing and street murals are common in North American Chicano popular culture (Simson, 1977); especially in Los Angeles (Cesoretti, 1975; D'Angelo, 1976; Kim, 1998; Romotsky & Romotsky, 1974, 1977); and in Texas (Bohley, 1989; Franklin, 1995). Black graffiti has also served a special ethnic purpose (Jefferson, 1976), and many of the civil rights slogans were messages on graffiti-like pickets; and exploitative ghetto billboards for liquor and tobacco have been defaced in Black ghetto protests.

A related adaptive blend of graffiti in cultural cooperation, compromise and co-optation is that of its combination with rap music and break dancing (Hager, 1984). It was "transplanted intact, embracing language, history, customs and rules..." (Chalfant & Prigoff, 1987, p. 8), somewhat like graffiti in the 1960s ushered in "the first genuine teenage street culture since the fifties" (Goldstein, 1973, p. 39). The cultural fabric is woven with colorful language, music, and clothing enveloping rural, ethnic and native youth in a new urban setting. For Blacks and Puerto Ricans, "Their common experience of racist exclusion and social distance from their white-ethnic peers drew them even closer together" through rap, graffiti and break-dancing (Flores, 1987, p. 580); although the first major subway graffiti writer was Taki, a Greek-American.

A kind of social organization has developed among hip-hop graffiti writers through crews and networks in contact with other writers' groups; classes for mentors and proteges; and multi-colored, mural-like tags of markers, throw-ups and pieces on public walls. Most participants are Black and Latino inner city young males seeking interaction as well as exposure of their artistic fame, a degree of claim and control over a city area; and a chance to rebel and protest against conventional norms (Brewer & Miller, 1990). Graffiti is often used by marginalized members of society to deal with tensions about gender, sexual orientation, race and social identity, as it provides "an anonymous forum for people to speak about issues that are often sequestered in a way that is usually frowned upon" (Rodriguez & Clair, 1999, p. 13). It can move beyond walls to tense sidewalk talk and encounters in a form of "interactional vandalism" (Duneier & Molotch, 1999); and even into the mainstream through "guerrilla marketing" whereby street teams use surprise, interactive, word of mouth, foot-based, two-way, grass-roots level communication to promote company products with fliers and free samples (Hays, 1999) in direct confrontation and even more so with new social and interactive media.

Whether in art or action, graffiti writers "resist the increasing segregation and control of urban environments" and the writer not only "confronts and resists existing arrangements but constructs alternative social, cultural and economic arrangements as well." (Ferrell, 1995, p. 73). "And like graffiti writing, these various movements of youthful resistance—too often dismissed as mindlessly destructive—in fact merit our attention not only for undermining contemporary social arrangements but for imaging new ones as well" (Ferrell, 1995, p. 87). New spatial boundaries are also created.

As "crimes of style", it is suggested that "Graffiti writing survives as a creative, playful response to a prefabricated culture, a response as disrespectful of legal and political boundaries as is the private and 'public' property through which they are enforced. Graffiti exists as a public art outside the control of public officials, an alternative style outside the circle of corporate style and consumption. Graffiti illuminates the city, but sporadically, less a series of lasting monuments than evocative moments, vulnerable to the give and take of the street. Ultimately, it stands as a sort of decentralized and decentered insubordination, a mysterious resistance to conformity and control, a stylish counterpunch to the belly of authority" (Ferrell, 1993, p. 197).

Conclusions

> Graffiti is perhaps the exemplary mode of outlaw communication. By definition, it is intrusive, emblematic, and opportunistic. ... Common to all types of graffiti is that they are written on what is taken publicly to be an illegitimate canvas. Graffiti is always an intrusion, and in this sense it is premeditatedly –but purposefully—*out of place.*
>
> (Emphasis in original) ... Graffiti is also a symptom of the failure or the limits of permissible or legitimate communication within the public sphere ... challenges us to think about the relationship between people and places ... a profoundly iterative form of urban expression ... as a communicative technology is crucially implicated as an emergent spatial politics of entitlement and belonging within the city ... alternative stories about places ... signal the failure of the public sphere to incorporate them ... graffiti has endured as the most archaic form of written communication. In this sense, we need to be able to read the writing on the wall, for in these opaque and arcane inscriptions, we can find both the limits of public discourse and the outlaw voices that threaten to be heard. (Back et al., 1999, pp. 69–70, 97, 99)

As more and more words cross and trespass old boundaries, people see less and less of each other during computerized impersonal communication and "almost every writer on graffiti remarks on the fact that practically no one has ever been observed" in the act of writing graffiti (Gadpaille,1971, p. 74). Perhaps this suggests new non-geographic and ghostly roles for graffiti perpetrators and legal prosecutors debating whether graffiti is art, a crime, or just urban communication clashing in debates on the street in a situational art gallery where entitled property owners aided by governments and the courts perpetuate and guard their territory from the anger and danger of strangers seeking to construct a new "architecture" for communication to invade and adapt city space for their own ends through new forms and spaces for urban citizenship. That cities may and will continue to attempt to regulate graffiti seems likely. Ironically, such regulation is important both to the property owners and to those wishing to break the law with their urban graffiti.

References

Abel, E. L., & Buckley, B. E. (1977). *The handwriting on the wall: Toward a sociology and psychology of graffiti.* Westport, CT: Greenwood.

Abrams, C. (1971). *The language of cities.* New York, NY: Equinox.

Back, L., Keith, M., & Solomos, J. (1999). Reading the writing on the wall. In D. Slayden & R. K. Whillock (Eds.), *Soundbite culture—The death of discourse in a wired world* (pp. 69–102). Thousand Oaks, CA: Sage.

Bayer, M. (1980) Saying goodbye through graffiti. *American Journal of Nursing, 80*(27).

Becker, H. (1963). *Outsiders: Studies in the sociology of deviance.* New York, NY: Free Press.

Bennett, P., Kendall, A., & McDougall, J. (2011). *After the media.* London: Routledge.

Bohley, S. (1989). *An anthology of the spray can: Chicano graffiti in Austin, Texas* (Master's Thesis). University of Texas, Austin, TX.

Brewer, D., & Miller, M. (1990). Bombing and burning: The social organization and values of hip hop graffiti writers and implications for policy. *Deviant Behavior, 11*(4), 345–369.

Brighenti, A. M. (2010). At the wall: Graffiti writers, urban territoriality, and the public domain. *Space and Culture, 13*(3), 315–332.

Brooks, J. (1989, June 16). Conservators at Ellis Island find multilingual graffiti. *New York Times II, 4*, p. 3.

Brown, J. (1995). The writing on the wall: The messages in Hungarian graffiti. *Journal of Popular Culture (Fall), 29*(2), 115–118.

Bushnell, J. (1990). *Moscow graffiti: Language & subculture.* Boston, MA: Unwin Hyman.

Carrington, K. (1989). Girls and graffiti. *Cultural Studies, 3*, 89–100.

Cesoretti, G. (1975). *Street writers: A guided tour of Chicano graffiti.* Los Angeles, CA: Acrobat Books.

Chaffee, L. (1989). Political graffiti and wall painting in Greater Buenos Aires: An alternative communication system. *Studies in Latin American Popular Culture, 8*, 37–60.

Chalfant, H., & Prigoff, J. (1987). *Spray-can Art.* London: Thames & Hudson.

Clay, G. (1987). *Right before your eyes—Penetrating the urban environment.* Washington, DC: American Planning Association.

Davies, J. (2012). Art Crimes?: Theoretical Perspectives on Copyright Protection for Illegally Created Graffiti Art 65 Me. L. Rev. 27 http://digitalcommons.mainelaw.maine.edu/mlr/vol65/iss1/3

D'Angelo, F. (1976). Fools' names and fools' faces are always seen in public places. *Journal of Popular Culture,* 10(1), 102–109.

Duneier, M., & Molotch, H. (1999). Talking city trouble: Interactional vandalism, social inequality and the 'urban interaction' problem. *American Journal of Sociology,* 104(5), 1263–1295.

Feiner, J., & Klein, S. M. (1982, Winter). Graffiti talks. *Social Policy, 12,* 47–53.

Ferrell, J. (1993). *Crimes of style: Urban graffiti and the politics of criminality.* New York, NY: Garland.

Ferrell, J. (1995). Urban graffiti: Crime, control, and resistance. *Youth & Society, 27*(1), 73–92.

Fleming, R. L., & von Tscharner, R. (1981). *Place makers: Public art that tells you where you are.* New York, NY: The Townscape Institute & Hastings House.

Flores, J. (1987). Rappin, writin and breakin. *Dissent* (Special Issue—Fall), *34*(4), 580–585. In Search of New York.

Franklin, O. P. (1995). *Listen to the walls speak: Murals and symbology in Austin, Texas 1990–1995* (Master's Thesis). University of Texas, Austin, TX.

Gadpaille, W. J. (1971, November). Graffiti: Psychodynamic implications. *Sexual Behavior, 2,* 45–51.

Glazer, N. (1979, Winter). On subway graffiti in New York. *Public Interest, 54,* 3–11.

Goldstein, R. (1973, March 26). This thing has gotten completely out of hand. *New York Magazine, 6*(13), 35–39.

Grieb, K. J. (1984, Summer). The writing on the walls: Graffiti as government propaganda in Mexico. *Journal of Popular Culture, 18,* 178–191.

Gross, D. (1993, Fall). Tagging: Changing visual patterns and the rhetorical implications of a new form of graffiti. *Etc, 50*(3), 250–265.

Hager, S. (1984). *Hip hop: The illustrated history of break dancing, rap music and graffiti.* New York, NY: St. Martin's.

Hall, E. T. (1959). *The silent language.* Garden City, NY: Doubleday.

Hall, E. T. (1969). *The hidden dimension.* Garden City, NY: Doubleday.

Hays, C. L. (1999, October 7). Guerilla marketing is going mainstream. *New York Times,* pp. Cl, 8.

Horn, J. (1976, February). Gang graffiti-spray-can boundaries. *Psychology Today,* pp. 100–102.

Hougan, J. (1972). Kilroy's new message: Political graffiti. *Harper's, 245,* 20–25.

Huber, J. (1986). *Paris graffiti.* London: Thames & Hudson.

Jacobs, K., & Heller, S. (1992). *Angry graphics: Protest posters of the Reagan/Bush era.* Salt Lake City, UT: Peregrine Smith.

Jefferson, R. S. (1976). Black graffiti: Image and implications. *Black Scholar, 7,* 11–19.

Keith, M. (2005). After the Cosmoplitian? Multicultural cities and the future of racism, New York: Routledge, p. 152.

Kim, S. (1998). *Vital signs: Signage, graffiti, murals and sense of place in Los Angeles* (PhD Dissertation, UCLA, 1997).

Klinghoffer, D. (1993, August 23). Wall-eyed. *National Review, 45*(16), 63–64.

Koleilat, N. R. (1985). *War graffiti in Beirut, 1975–1982: As religious, political and social communication in Lebanon* (Master's Thesis). University of Texas at Austin.

Lachmann, R. (1988). Graffiti as career and ideology. *American Journal of Sociology, 94*(2), 229–250.

Landy, E. E., & Steele, J. M. (1967). Graffiti as a function of building utilization. *Perceptual and Motor Skills, 25*, 711–712.

Ley, D., & Cybriwski, R. (1974). Urban graffiti: Territoriality in an urban context. *Annals of the Association of American Geographers, 64*, 491–505.

Lindsay, J. (1960). *The writing on the wall: An account of Pompeii in its last days.* London: Frederick Mueller.

Mailer, N. (1974, May). The faith of graffiti. *Esquire,* pp. 154–158.

Mockridge, N. (1968). *The scrawl of the wild: What people write on walls and why.* Cleveland, OH: World Publishing.

New York Magazine. (1973, March 26). The graffiti 'hit' parade, pp. 32–35.

New York Times. (1999, April 9). Lawyer likens Melissa virus to graffiti, p. A23.

Nicholson, S. (1972, October). The theory of loose parts. *Landscape Architecture.*

Nilsen, D. L. (1980). Grammar of graffiti. *American Speech, 55*(3), 234–239.

Olivero,L.(2014,July11).GraffitiIsaPublicGood,EvenAsItChallengestheLaw.NewYorkTimes, https://www.nytimes.com/roomfordebate/2014/07/11/when-does-graffiti-become-art/graffiti-is-a-public-good-even-as-it-challenges-the-law?mcubz=3

Opler, M. K. Graffiti represent thwarted interests. In W. J. Gadpaille, *Graffiti.*

Powers, L. A. (1996, Spring). Whatever happened to the graffiti art movement? *Journal of Popular Culture, 29*(4), 137–142.

Raymond, C. (1989). Scholar finds arts, social tradition and graffiti that many dismiss as chaotic scribbling. *Chronicle of Higher Education, 36*, 1, 4.

Rhyne, L. D., & Ullmann, L. P. (1972). Graffiti: A nonreactive measure. *Psychological Record, 22*, 157–168.

Rice, L. (1991). *Up against it: Photographs of the Berlin wall* (Photo Collection). Austin, TX: Harry Ransom Center, University of Texas.

Rodriguez, A., & Clair, R. P. (1999, Fall). Graffiti as communication—Exploring the discursive tensions of anonymous texts. *Southern Communication Journal, 65*(1), 1–15.

Rolston, B. (1987). Politics, painting and popular culture: The political wall murals of Northern Ireland. *Media Culture & Society, 9*, 5–28.

Romotsky, J., & Romotsky, S. (1974). California street scene: Wall writing in Los Angeles. *Print, 28*, 60–65.

Romotsky, J., & Romotsky, S. (1977). L. A. Human Scale: Street Art of Los Angeles. *Journal of Popular Culture, 10*(3), 653–667.

Science Digest. (1969). Walls remember, *65*(6), 31–33

Science Digest. (1974). Graffiti helps mental patients, *75*(4), 47–48.

Sennett, S. (1990, Summer). New York reflections. *Raritan: A Quarterly Review 10*(1), 1–23.

Simson, E. (1977, Winter). Chicano street murals. *Journal of Popular Culture, 10*(3), 642–652; and "a sociological perspective" by Simson in 1980, 13(3), 516–526.

Solomos, H., & Yager, H. (1975). Authoritarianism and graffiti. *Journal of Social Psychology, 97*(1), 149–150.

Stocks, T., Dutcher, L. W., Hargrove, S. M., & Cook, E. A. (1972). Social analysis of graffiti. *Journal of American Folklore, 85*, 356–366.

Tillman, T. (1990). *The writings on the wall: Peace at the Berlin wall.* Santa Monica, CA: 22/7 Publishing.

Wales, E., & Brewer, B. (1976). Graffiti in the 1970's. *Journal of Social Psychology, 99*(1), 115–123.

Wikipedia. (2015). Graffiti in the United States. Retrieved from https://en.wikipedia.org/ wiki/ Graffiti_in_the_United_States

Young, A. (2013). *Street art, public city: Law, crime and the urban imagination.* New York, NY: Routledge.

CHAPTER FOUR

Urban Sign Regulation

HARVEY JASSEM
Professor Emeritus, University of Hartford

Cities are full of it. That's what makes them all the things they are—interesting, fun, handy, congested, maddening, home, tourist magnets, dangerous, safe, efficient, hard to categorize. And perhaps because of all of the cacophony that is "urban," one role played by city governments is that of arbitrator of that oh-so-shared space. This chapter looks at the ways in which cities attempt to place "signs" in their proper place in that space. Signs, whose very purpose is to be noticed and to communicate, often with those not intending to be communicated with, are as complex as cities themselves. They come in all shapes and sizes, are useful and distracting, loved and hated, and both regulated and protected from regulation.

Signs are important. We would be lost without them. They assist us with direction finding, even in a GPS era. They show addresses, street names, and tell us where to turn for the nearest beach, mall, school, church, or whatever. They inform us about what is around us. It could be a hardware store, post office, gas station, hospital, or, again, whatever. Signs tell us about our environment. The road is closed ahead, the temperature and time are …, this is a "quiet" zone, blind person crossing, George Washington slept here. They can be informative or persuasive, commercial or non-commercial, political, offensive, entertaining, crude, beautiful, or all or none of these things. And that is just describing their use. There is perhaps an even broader characterization of their form, which will be discussed later.

Signs are important communication tools and as such they are protected by the First Amendment.

And signs are *intentional* communication tools. They are intrusive by design. They are not designed to be ignored. They are designed to stand out from their environment and be seen. That is not to suggest that good sign design requires thumbing one's nose at the context, but it does mean that if a sign blends in so well as not to be noticed, it is not an effective sign for most contexts. As such, signs change their environment. They block what is behind them, they take attention away from other things, they may physically get in the way, their content may grab and shape people's attention. They are there to make a difference, and the intentionality behind the sign suggests that the difference sought is often in the interest of the entity or person responsible for the sign.

As we consider a sign in the city, it is important to remember that predictably, as one tries to change the environment through signage, there are likely to be conflicting interests, and that is where the regulation of signs comes to play. Governments have many interests that bump up against the communication medium of signage. Governments have an interest in the safety of people, in the safe and efficient functioning of the city, in aesthetics and the economic well being of the city. All of those, and some others, lead cities to regulate signs.

And given the interests of those wanting to communicate (or be communicated with) through signs, those regulations have been challenged.

This chapter looks at why signage is regulated, the process and goals of sign regulation, and problems and difficulties in developing and applying sign regulation especially in light of recent court decisions. A review of the state of urban sign regulation as well as suggestions for those wanting to regulate signs in cities will be offered. For purposes of this review, the focus of this chapter is on "signage" that is not authored or created by the city government itself. Hence, for example, a city's rules regarding how it handles its own street signs is not discussed.

Why Regulate Signs?

Signs are regulated as part of a municipality's authority to zone and control public spaces in order to achieve and protect public safety and aesthetics, especially as aesthetic concerns could reasonably impact quality of life, property values, and even economic health of a municipality.

There are, of course, good reasons to regulate signs, and municipalities have been regulating signs for over a century. Signs can in themselves be unsafe, fall down and hurt people, block traffic lights or the ability to see oncoming traffic, create all sorts of health and safety hazards, and be incompatible with the neighborhood they're in. They can also be a public nuisance as they impact a community's sense of self and economic well-being. Alan Weinstein, former Chair of the American Bar Association's Subcommittee on Land Use and the First Amendment, reported for the American Planning Association (Morris, Hishaw, Mace, & Weinstein, 2001, ch. 6; Weinstein, 2001) that municipalities have regulated signs for over 100 years as part of their "police power" to protect public safety and preservation of property values. These concerns have expanded over the years to include aesthetics and economic well being of the city.

Public safety concerns are pretty straightforward. Regulations that are designed to protect people from structurally defective signage, from signs that block traffic lights, and even from signs that might be easily confused for traffic lights and thus cause traffic accidents, if narrowly and neutrally tailored, are easy to justify. More controversial are regulations that look to address issues of aesthetics, as those impact quality of life and the economic repercussions. In these cases, the issues tend to be tied to the notion of zoning. There's a time and place for things. A big bright ugly sign in a quaint charming neighborhood might just ruin the quality of life and the property and economic values of that neighborhood. And while such regulation may seem quite justified to those feeling adversely affected by such signage, the justification of such regulation is more rigorously tested when First Amendment concerns are brought to the fore.

A great deal of case law exists regarding the application of First amendment protections around sign regulation. When a state (or, in this case, city) action restricts the content or expression of speech, the state generally has the burden of proof to demonstrate why such action is necessary. Such restrictions must generally pass the "strict scrutiny" level of review—that is, the restriction must very narrowly address a compelling government interest and use the least restrictive means possible to achieve those interests. Restrictions on purely commercial speech and restrictions that are form (or "time, place, manner"—"TPM")-based, rather than content based, often require the government to meet "intermediate scrutiny" tests. These are easier tests for the government to meet. The government would have to show that the restrictions are very narrow, address an important or substantial government interest (that is unrelated to the content of the communication) and still leave ample opportunity for alternative means of communication such messages (Central Hudson, 1980; Ladue, 1994).

Sign regulation, as is the case with much speech regulation, requires some balance and contextual analysis, despite the absolutist-sounding dictates of the First Amendment ("Congress shall make *no* law …"). Regulations pertaining to language requirements in signs present one easy case study. Some U.S. City governments have been concerned about the prevalence of non-English signage in some neighborhoods, and those governments have sometimes attempted to require

English in all signage. (In parts of Canada, regulations *prohibit* the use of English in some signage.) Is a language requirement a content regulation, is it cultural discrimination, is it about public safety? In the U.S., when such concerns are about cultural purity, integration, and the like, they tend not to be codified as regulation because regulating speech targeting such concerns would not pass Constitutional muster. When, on the other hand, there are serious public safety reasons behind such regulations (perhaps a requirement that an "exit" sign be in English or in English and another prevailing language, or that warning signs relied on by police or fire departments be in English), such regulations could withstand judicial review. So, it isn't that language or content cannot be regulated, but such regulations must be narrowly tailored to address important and legitimate government concerns.

The Process/Home of Sign Regulation

Urban sign regulation is generally tied to both zoning and building review functions of municipal governments. Permits are often required in order to erect a sign. The permitting process, especially because signs tend to be about communication, cannot be required in a discriminatory manner nor used as an unreasonable form of prior restraint. If it is illegal to regulate based on some sort of content, for example, it would be illegal to require permits based on that content. Permit requests must be handled in a timely fashion. Those permits may address both quantitative and qualitative issues, as well as placement issues. Quantitative issues include such things as: what are the dimensions of the sign, what are the construction materials, is the sign lit, and if so by what means, how tall is the sign, what are the clearances around the sign or between the sign and nearby rights of way/buildings/other signs, etc.? While such issues might seem to be straight forward and not subject to interpretation, that is not always the case. If a sign is of an irregular shape, where does one measure from? If the sign is projected onto a building or a sidewalk, and its size varies with the changing graphic, how big is that? If the sign is attached to a bracket that looks like the company logo, is the bracket part of the sign, or is it a bracket? And, as will be discussed later, there might even be a question regarding what actually constitutes a "sign." Still, these quantitative elements are the easier aspects to describe.

Qualitative issues can be more prone to subjective interpretation. In some areas, signage must conform to the aesthetic sensibility of the neighborhood. Perhaps it is a quaint historic district with regulations requiring design review for buildings and signs in order that they maintain the historic look of the area. Perhaps there is a question about the distractive nature of the sign, or a question about whether the sign enhances or detracts from the surrounding location. There might even be questions about the intent of the sign content. All of these kinds of concerns are judgment issues, raising serious First Amendment threats, yet all are frequently invoked and have often been upheld in the courts.

City sign regulations vary by location. New York City's Department of Buildings (DOB, n.d.) summarizes it this way:

> Many aspects of a sign (height, size, illumination etc.) are governed by its zoning district and location. In New York City, there are three types of zoning districts
>
> - Residential
> - Commercial
> - Manufacturing
>
> Each District has different rules and regulations when it comes to signs. For example, Times Square has unique sign requirements (big, bold and bright) that do not apply anywhere else. Residential districts have the most restrictive sign regulations.

In New York City residential districts, advertising signs are generally prohibited, with some exceptions made for "accessory signs" (essentially signs that are attached to and about the business to which they are attached—these are often known as "on premise" signs) and "for rent" or "for sale" signs. Illuminated signs are only permitted for hospitals, are not permitted to flash, "must be limited to the lesser of 25 square feet per street frontage or 15 percent of such street frontage," and must be positioned such that they minimize the light projected into residences (Thompson, 2012).

In New York City commercial districts, the regulations are far more complex for accessory signs, and advertising signs are allowed in some, though not all, commercial districts. In manufacturing districts, the sign regulations are less restrictive still. Even in these districts, there are restrictions that deal with safety and quality of life for those nearby.

Before erecting a sign, municipalities generally have sign permitting processes that applicants must adhere to so that intended signs may undergo review before being constructed. The cost and complexity of the permitting process (does one need licensed architects' drawing or specs, does one need environmental impact statements, etc.?) vary by municipality and by the type or scope of sign one is proposing. The permitting process should work to ensure that sign regulation is done fairly and publically, and can provide public records relating to the nature of the approval. Those records could be useful in later proceedings.

While there is some federal sign regulation, most notably the Highway Beautification Act (1965), the vast majority of sign regulation, and that which is the focus of this chapter, is locally based. It is generally found in zoning regulations, but may also be in departments of transportation, public safety, buildings, or public transit.

The Sign Regulations

As pointed to earlier, sign regulations deal with content, size, color, placement and a myriad of issues that are in themselves interesting and complicated. First, and foremost, however, is the question of just what is a "sign." That is not as straightforward as it might seem.

A New York State publication designed to help cities and towns craft sign regulation notes:

> Definitions are a key component of sign regulations since many municipalities regulate some signs differently. Municipalities wishing to regulate both traditional and non-traditional signs should provide accordingly. Well-defined terms can also help facilitate the duties of the municipal boards and enforcement personnel involved with administering the regulations. Some municipalities include illustrations to support the written definitions. The Town of Huntington regulates signs through its zoning regulations … In its zoning regulations, Huntington broadly defines "sign" as: Any structure or part thereof, or any device or group of letters attached to, painted on or represented on a building, fence or other structure on or in a window or temporarily or permanently on a vehicle or trailer, upon which is displayed or included any letter, symbol, trademark, model, banner, flag, pennant, insignia, decoration, device or representation used as or which is in the nature of an announcement, direction, advertisement or other attention-directing device. A "sign" does not include the flag or pennant or insignia of any nation or association of nations or of any state, city or other political unit or of any charitable, educational, philanthropic, civic or religious organization. (New York State Department of State, 2006, 2015, p. 31, citing Huntington Code §198–2)

New York State's town of Brookhaven defines a "sign" as:

> Any material, structure or device or part thereof composed of lettered or pictorial matter or upon which lettered or pictorial matter is placed when used or located out of doors or outside or on the exterior of any building, including window display area, for display of an advertisement, announcement, notice, directional matter or name, and includes sign frames, billboards, signboards, painted wall signs, hanging signs, illuminated signs, pennants, fluttering devices, projecting signs or ground signs, and shall also include any announcement, declaration, demonstration, display, illustration or insignia used to advertise or promote the interests of any person or business when the same is placed in view of the general public. (NYSD, 2006, 2015, p. 33)

Other towns and cities have their own definitions. And while there is some overlap, there are clearly differences. Just look at the two examples above; there are a number of differences. Other definitions of signs sometimes include identifiable architecture, identifiable colors, special lighting, and so forth. Beyond that, sign regulations tend to call out subsets or categories of signs, including such things as wall signs, canopy signs, roof signs, projecting signs, pole signs, flags, suspended signs, window signs, letters, sky signs, neon signs, barber poles, temporary signs, "for sale" signs, movable signs, signs with changing content, etc.

That the definitions used by cities and towns are inconsistent presents a problem for sign makers, planners, and the courts. But even if they were all the same, there are still potential problems defining "sign." Some examples:

What is a truck that has a message painted on it? Perhaps that message is the name of the company owning and operating the truck. Perhaps the message is the logo or an ad for a product. Maybe it's a political statement. Is the truck a truck or is it a sign? Does it matter if the truck is moving on the public road or parked on that road? How about if that truck is parked on private property near the road? Does it make a difference if the private property is owned by the truck's owner operator? If so, why? Would it make a difference if that truck box were filled or empty? And if not, what is a flatbed truck that has a billboard mounted on it … a truck or a sign?

What is a pile of tires? On the one hand, a pile of tires is a pile of tires, and tires aren't signs. How about if that pile of tires is in the front lot of a tire store? Should that pile of tires be regulated differently than the pile of tires next door, when that next door's pile of tires simply reflects a messy owner rather than an establishment that's selling tires?

Similar questions have been raised about a national flag. National flags are generally exempted from sign codes. But what of an Irish flag on the outside of an Irish-themed bar? Some have argued successfully that in that case, the national flag is a sign, even if the same flag would not be considered a sign someplace else down the street.

How about a sculpture that is large and calls attention to itself. Is that a sign? A proposal for a 32 foot-high cylindrical structure at the entrance to a shopping plaza was denied by a town because the zoning commission found that "the structure's purpose is to attract attention, therefore it is indeed a sign" and did not meet the sign zoning regulations. Connecticut courts ultimately struck down this finding that the town's "expansive construction of the term 'sign' … to include structures or objects that are clearly sculptures or works of art, is unsupported by" the state's statutes that enable local zoning.[1] Notable here is that the court found the statutes didn't apply to things that were "clearly sculptures or works of art." Clear to whom is not itself clear. Would the Apple window display of hundreds of multi-colored little LEDs that at once permit an unobstructed view into the store while simultaneously changing design be a sign or art? What happens when one of those abstract designs morphs briefly into the Apple trademarked logo of an Apple? The court's finding in this Connecticut mall art case was not that such works could not be deemed to be signs in any situation, but that the Connecticut state statutes did not envision that such art could be subject to zoning commissions' sign regulation. Perhaps other statutes or other states' statutes might be written or interpreted differently.

Conversely, could what is clearly and obviously a sign sometimes be art and not a sign at all? A clothing store in a hip Tribeca neighborhood took over a storefront that had previously been a liquor store. The only signs outside the clothing store and stenciled onto the windows were from and about the liquor store. They were old cool looking signs, but said nothing about the clothing. They may have implicitly been suggesting nostalgia and change, but can implicit meaning be regulated? The only indication that it was a clothing store were the racks of clothing that could be seen through the windows. Were those racks of clothing the sign?

Also in New York City, there is a huge iconic Pepsi Cola sign on the East River facing Manhattan. The *New York Times* wrote:

> Once regarded as an eyesore, the sign is generally embraced today as a symbol of Long Island City's industrial past, as a colossal work of Pop Art …
>
> It seems that if you keep a billboard long enough, it may turn into a civic cynosure. The Citgo sign at Kenmore Square, for instance, probably ranks ahead of the Old State House as a symbol of Boston. And the Pepsi sign was once considered for landmark status.[2]

Figure 4.1. Neon Pepsi—is it art or is it a sign?

The Pepsi sign once stood atop its factory in an industrial district. Now that factory is gone and there are residential buildings going up. It is no longer an on-premise sign announcing what's in that building. Might such a sign no longer be a sign and take on a new character as some sort of art? That is not simply a rhetorical question. Many grandfathered signs are subject to the new regulations if they need to be rebuilt after facing extensive damage. That Pepsi sign is just about at ground level adjacent to the East River which some predict will be prone to severe flooding as a result of climate change. Would that Pepsi sign, which is about as clearly a sign as one could imagine, be required to meet new signage rules if destroyed, or would it be considered art and thus not subject to those rules? And if that beloved sign is the latter, what if Pepsi tried to erect its duplicate elsewhere in the City in zones that don't permit billboards or signs? Would it be permitted because it is considered to be "art?"

And what if a wall-sized mural contained images that include images that are part of a company's logo? What if a mural contains "political" content? Are they murals (art) or are they signs? Could their permissibility be impacted by changing some aspect of the sign/mural as municipalities have sometimes suggested? (Weinstein & Connolly, 2015, pp. 23–26).

Of course, some of these questions are addressed in the ordinances, and all ordinances are given meaning through judicial review of their application, since no ordinance can foresee every single possible application. But it is important to note that even before we delineate sign regulations we must acknowledge that even the definition of "sign" is less than clear.

Categories of Signs

Because signs tend to be regulated in conjunction with zoning, and zoning is largely about place, much of sign regulation takes into consideration place. As noted above, there tend to be different regulations for different kinds of neighborhoods. But even within any particular neighborhood or zone, regulations differ by "type" of sign. Steiner and Butler (2007, p. 375), for The American Planning Association, summarize the type of permissible sign regulations as follows.

Sign characteristic	Permissible regulations
Dimensions	Maximum size, height, degree of projection, measurement of sign area, number per lot/building
Type	Freestanding, monument, pole, wall, projecting, portable, temporary, electronic reader boards, and so on

(Continued)

(*Continued*)

Sign characteristic	Permissible regulations
Location	Setbacks from street, sidewalk, placement on buildings, prohibited areas, rules for corner lots
Lighting & Illumination	Allow or prohibit internal lighting, external lighting, flashing signs, neon, electronic reader by animation or neon
Sign materials/colors	Allow or prohibit cabinet signs, wood signs, metals, colors, gold leaf

The reader should keep in mind that this represents a very brief summary of the sign characteristics regulations address. A complete list of other delineated sign characteristics could literally fill this book. While such specificity might be warranted, regulators must be careful not to design rules that are clearly targeting a particular sign in ways that make it, and not other signs, subject to removal or regulation.

Once sign categories are identified, and once the zone is identified, regulations may identify what is permitted when. The "what" here is largely not content-based, although even that is arguable (Might the iconic color of a business be considered "content"?). Beyond regulating sign characteristics, content itself is also sometimes the focus of regulation. The legality of all such restrictions are subject to judicial review.

Legal Review

One might imagine legal cases about big and literally flashing signs. But what of smaller less obtrusive situations? What if, for example, a religious group wanted to post small, non-illuminated temporary signs directing worshipers to their weekly services, and such signage was necessary because the services were held in different locations weekly. That is just the sort of thing that the town of Gilbert, Arizona, tried to stop by its application and enforcement of its sign regulations, and it is the most recent significant Supreme Court case dealing with sign regulation.

Reed v. Gilbert

Gilbert's 2005 sign code prohibits all outdoor signs without a permit, but exempts 23 categories of signs. In its decision in this case, the U. S. Supreme Court notes that three of those exempted categories include

- Ideological signs for noncommercial purposes. They are afforded the most protection—they may be "up to 20 square feet in area and … placed in all 'zoning districts' without time limits."

- Political signs "designed to influence the outcome of an election." They may be up to 16 or 32 square feet in area, depending on the type of property they're on, and may be displayed for a limited number of days prior to and following the election.
- Temporary directional signs relating to a qualifying event, where a qualifying event is defined as any "assembly, gathering, activity, or meeting sponsored, arranged, or promoted by a religious, charitable, community service, educational, or other similar non-profit organization." Such signs could be no larger than 6 square feet and may be displayed "no more than 12 hours before the 'qualifying event' and no more than 1 hour afterward." (Sec. 1 A, pp. 2225–2226)

The religious organization, Good News Community Church, and its pastor Clyde Reed, were cited by the town for posting directional signs to the church's services. The town asserted that the signs, some of which were posted on Saturday for Sunday services, exceeded the 12 hour limit and did not include the date of the event. The Church argued that this enforcement violated their First and Fourteenth Amendment rights. The District Court and Court of Appeals for the Ninth Circuit found in favor of the town. The Court of Appeals ruled that the regulations were "content neutral," even though an enforcement officer would obviously have to read the sign in order to know what regulations might apply and to determine if the sign was in compliance with the regulations. Such was "not akin to an officer synthesizing the expressive content of the sign." 587 F.3d 966, 978 (2009), but ordered the District Court to review the sign categories themselves to determine if they were content neutral. The District Court indeed reaffirmed its opinion that the sign regulations were content neutral, and as such did not warrant strict scrutiny standards of review. The US Supreme Court granted certiorari in 2014 and reversed the lower court's decision in 2015.

The Supreme Court found:

> The restrictions in the Sign Code … depend entirely on the communicative intent of the sign. If a sign informs its reader of the time and place a book club will discuss John Locke's Two Treatises of Government, that sign will be treated differently from a sign expressing the view that one should vote for one of Locke's followers in an upcoming election, and both signs will be treated differently from a sign expressing an ideological view rooted in Locke's theory of government. More to the point, the Church's signs inviting people to attend its worship services are treated differently from signs conveying other types of ideas. On its face, the Sign Code is a content-based regulation of speech. We thus have no need to consider the government's justifications or purposes for enacting the Code to determine whether it is subject to strict scrutiny. (2227)

The Court goes on to reassert its previous findings that regulations will be found to be content-focused, and not content neutral, not only if they target a particular view, but even if they target "types" of speech (2230). A regulation that considers content considers content, regardless of how benign the regulator's intent. "That is why we have repeatedly considered whether a law is content neutral on its face *before* turning to the law's justification or purpose." (2228) And the Court uses a broad definition of content regulation. It explains: "for example, a law banning the use of sound trucks for political speech—and only political speech—would be a content-based regulation, even it if imposed no limits on the political viewpoints that could be expressed. ... The Town's Sign Code likewise singles out specific subject matter for different treatment, even if it does not target viewpoints within that subject matter. Ideological messages are given more favorable treatment than messages concerning a political candidate, which are themselves given more favorable treatment than messages announcing an assembly of like-minded individuals. That is a paradigmatic example of content-based discrimination." (2230)

Noting the town's obligation to meet the "strict scrutiny" test, "which requires the Government to prove that the restriction furthers a compelling interest and is narrowly tailored to achieve that interest," the Court concludes that "the Town cannot do so." The Town's asserted interests of aesthetics and public safety are not directly impacted by having different kinds of content regulated differently (2231). The imposition of the strict scrutiny to such content-based regulations "will not prevent governments from enacting effective sign laws. ... Not 'all distinctions' are subject to strict scrutiny, only *content-based* ones are. Laws that are *content neutral* are instead subject to lesser scrutiny." (2232, emphasis in original) Regulations that deal with sign size, materials, lighting, portability, and on public property, placement, can all be effective means of addressing the town's interests in content-neutral ways. "A sign ordinance narrowly tailored to the challenges of protecting the safety of pedestrians, drivers, and passengers—such as warning signs marking hazards on private property, signs directing traffic, or street numbers associated with private houses—well might survive strict scrutiny." [2232]

This 2015 U. S. Supreme Court majority decision was written by Justice Thomas and joined by Justices Roberts, Scalia, Kennedy, Alito, and Sotomayor. There were no dissents. However, the broad definition of content-based regulation triggering strict scrutiny was not universally supported. In three different concurring opinions, representing a total of six of the Justices, reservations to this broad brush application were raised. Justice Kagan, joined by Justices Ginsburg and Breyer sum up the concerns shared by many of the affirming Justices:

[T]here is no need to decide in this case whether strict scrutiny applies to every sign ordinance in every town across this country containing a subject-matter exemption.

> I suspect this Court and others will regret the majority's insistence today on answering that question in the affirmative. As the years go by, courts will discover that thousands of towns have such ordinances, many of them "entirely reasonable.". … And as the challenges to them mount, courts will have to invalidate one after the other. (This Court may soon find itself a veritable Supreme Board of Sign Review.) And courts will strike down those democratically enacted local laws even though no one—certainly not the majority—has ever explained why the vindication of First Amendment values requires that result. (2239)

Given the recency of this decision, the lack of a clear consensus regarding just what constitutes "content" and "content neutral" regulation, and, as Justice Kagan notes, a lot of local sign ordinances that might raise issues similar to the ones in *Gilbert,* it is likely that the courts will have to further clarify how broadly they will apply the principles outlined in this case. Regardless, while it appears that sign regulations that might be construed as even tangentially dealing with content are likely to be subject to strict scrutiny review and hence need to be quite narrow and directly targeting government interests, other sign regulations appear to remain protected and can continue to serve public good. It also appears that the long standing distinction between political and commercial speech protections may disappear for signage in the wake of *Reed.*

Post "Reed v. Gilbert"—What Is a City to Do?

As noted, the *Reed* decision makes content-based or content-related sign regulations much more difficult to justify. Weinstein and Connolly (2015) suggest there may be creative ways around that difficulty. They suggest, for example, restrictions on "for sale" signs are likely to be considered content-based, and thus subject to strict scrutiny, and hence should be replaced by regulations targeting or describing, for example, "a temporary sign placed on property which is actively marketed for sale …" (pp. 22–23) Regulations about "grand opening signs" might be replaced with ones for "temporary sign[s] placed within a parcel of property … which may be displayed for a period [of] days following the sale, lease or other conveyance of the parcel." (pp. 62–63)

Such workarounds, however cumbersome, are content neutral and hence not likely to be subject to strict scrutiny.

Clearly the risks that sign regulations will be challenged have increased since *Reed*, and cities would be wise to review their regulations and enforcement policies to minimize challenges by carefully reviewing the ways in which existing regulations may speak to content-related issues, and then cleaning them up. In many cases, the real interests and goals of such regulations may have little to do with the content of the signage, and content-related terms may have simply been inserted into regulations as a convenient way of handling the issue. In those cases, regulations and enforcement should be revised now to excise content references. Of course, there may be some areas in which cities feel the need to continue to have content-based sign regulations. Where that is the case, there is nothing wrong with that, so long as such regulations can withstand strict scrutiny.

Weinstein and Connolly (2015) urge that new or revised sign codes be written with "a strong, well-articulated purpose statement," planning studies, showing how signage fits with the city's goals, and employ empirical supporting justification. Cities and towns have a history of claiming that their interests in sign regulation are for such things as public safety and traffic administration, yet rarely produce substantial evidence (Jassem, 2014). Especially given the expected higher scrutiny of sign regulation, cities must make clear their interests and rationale.

All of this refers to sign regulations that, as most regulations do, attempt to limit signage. There is some growing interest in encouraging signage. There is some recognition that even more and bigger and brighter signage can help define districts or cities themselves. Regulations of that sort should be safe from judicial review if they are applied fairly and do not unduly infringe on other rights.

For example, New York City's Times Square sign regulations *require* bright illuminated signs. The intent of the regulations, according to Daniella Eidelberg of the Times Square Business Improvement district, was to "preserve the bright lights and signs that have made Times Square famous. … They help create the ambience of Times Square and distinguish it from other parts of Midtown." (Gerend, 2003). As Aaron Rutkoff (2010) noted in the *Wall Street Journal,* this doesn't simply mean that large bright signs are permitted or even encouraged. They are required! He writes: "Own a building on Broadway but detest the flashing lights? Too bad. As the code states:

> There shall be a minimum of one illuminated sign with a surface area of not less than 1,000 square feet for each 50 linear feet, or part thereof, of street frontage.

… Does your building feature a blinking sign? The rules require that the unlit phase not exceed three seconds. When can the bright lights be switched off? No earlier than 1:00 a.m."

Such regulations do not invoke any content concerns and should not be subject to strict scrutiny. They may survive.

Similarly, there is interest in finding ways to keep vintage signage that might not meet modern sign codes. As zoning and sign codes have pushed for rather standardized development and somewhat uniform looks, there are some who wish to maintain the quirky, and signs are part of the quirky. Of course old signs that don't comply with new codes are often grandfathered in, but generally only so long as they (or the property with which they are associated) remain substantially unchanged. Changes generally require that the new zoning and sign code provisions be applied. That does not always make for the best outcome.

In Tucson, Arizona, for example, there was a 1950s neon sign that incorporated a figure of a diving girl in a two piece bathing suit announcing the Pueblo Hotel. The sign had become something of a local landmark, even as it became rundown. When a law firm moved into and renovated the space, it wanted to put its name on a refurbished diving-girl neon sign, but the sign broke all sorts of modern signage requirements. Adding its name to the sign would not simply be a maintenance issue, it would be a significant change to the sign necessitating a permit, and that would raise problems because the sign would not meet modern sign codes. It was too large, didn't meet the required setback requirements, and failed to meet the size and place rules for pole and ground signs. The city denied the permit for change. That led to community and government meetings and the establishment of a "Historic Landmark Sign Code" of 2011. The code's stated purpose is, in part, "to provide for the preservation of the City of Tucson's unique character, history, and identity, as reflected in its historic and iconic signs, and to restore the sense of place that existed within the central business district and in areas of the City with concentrations of surviving historic signs." If signs meet this historic sign code's requirements, they are incentivized. They can be larger and brighter than otherwise would be permitted. And the code applies to historic signs, wherever they may be in the city, not just in designated districts (Testa, 2016).

The Supreme Court's *Reed v. Gilbert* decision shines a bright light on municipal sign regulation, and cities will have to be more careful about crafting their sign-like regulations so that they focus more on concerns of "time-place-manner" than on content if they don't want such sign regulation to be required to withstand strict scrutiny. But that will never be easy. Even this lovely vintage historic sign code may well find itself challenged on content grounds. What makes one old sign "historic" and another not? The answer often involves content.

And beyond content, there is another complication brewing for even TPM based regulations. Time, place, manner rules suggest that such things are rather stable and definable. They also suggest that those regulations apply within the jurisdiction within which the city government rules. Where are holograms, and how large and where are they? Where are signs that are projected onto large buildings, balloons, or even clouds? How bright are new technologies that emit light in ways that differ from the sort of light that old light meters were designed to measure? And what if we start walking through cities wearing virtual reality "glasses" that others post messages on as we traverse through the city. Are those "signs"?

Further, when are sign modifications minor and thus not necessitating permit and review, and what modifications trigger review? Modern signs complicate this already vague distinction. When signs were static and updating them might have meant new paint and wood, that's one thing. What about an electronic sign where the update might mean new forms of lighting or new computer controls which might impact the signs' ability to display movement, changing images, changing lighting intensities, etc. Sign codes may need to be amended to include attention to such details. In any case, they need to look again at how they attempt to define the difference between insignificant and significant modifications.

Such questions are limitless. And the regulation of signs will continue to be a necessary challenge. In the United States, sign regulations should be guided by explicated principle and work to protect both the freedom of expression and even freedom from expression. The permissible scope of sign regulation may change based on court directives, changing mores, and changes in the ways in and tools with which we live. There is not a single uniform sign code that should be universally adopted by cities. But, universally, all sign codes should be clear and focused on achieving the city's goals in as light handed a fashion as is possible. *Reed,* and changes in signage technology, may make it more difficult for cities to regulate signage, but the need for well crafted sign regulation remains.

Notes

1. Guilford Planning and Zoning Comms v. Guilford Zoning Bd of Appeals Conn Super Ct April 21, 2004, p. 3 of 8, citing Schwartz v. Planning and Zoning Commission, 208 Conn. 146 (1988).
2. David W. Dunlap, "Saving a Spot for Pepsi-Cola as a Tower Goes Up," *New York Times,* July 10, 2013, retrieved from http://www.nytimes.com/2013/07/11/nyregion/as-a-queens-tower-rises-a-spot-is-saved-for-pepsi-cola.html (March 18, 2014).

References

Central Hudson Gas & Electric v. Public Service Commission, 447 U.S. 557 (1980).

Department of Buildings (DOB). (n.d.). *About storefront sign requirements.*

Gerend, J. (2003). *The outrage over New York City's storefront awning ticket blitz is justified—but so are the limits.* Center for an Urban Future. Retrieved from https://nycfuture.org/research/publications/the-outrage-over-new-york-citys-storefront-awning-ticket-blitz-is-justified

Highway Beautification Act, as amended. 1965. 23 USC 131.

Hinshaw, M. (2001). Design review, permitting, and recordkeeping. In M. Morris et al. (Eds.). *Context-sensitive signage design.* American Planning Association Planning Advisory

Service. Retrieved from https://planning-org-uploaded-media.s3.amazonaws.com/legacy_resources/research/signs/pdf/contextsensitivesignagedesign.pdf

Jassem, H. (2014). *Bad sign*. Presented to Eastern Communication Association, Providence, RI. Available from author.

Ladue v. Gilleo, 512 U.S. 43 (1994).

Morris, M., Hishaw, M. L., Mace, D., & Weinstein, A. (Eds.). (2001). *Context-sensitive signage design*. American Planning Association Planning Advisory Service. Retrieved from https://planning-org-uploaded-media.s3.amazonaws.com/legacy_resources/research/signs/pdf/contextsensitivesignagedesign.pdf

New York State Department of State, *Municipal Control of Signs*. (2006, revised 2015). Retrieved from https://www.dos.ny.gov/lg/.../Municipal_Control_of_Signs.pdf

Reed v. Gilbert. 135 S.Ct. 2218 (2015).

Rutkoff, A. (2010, August 12). Good taste in times square? It's illegal. *Wall Street Journal*. Retrieved from http://blogs.wsj.com/metropolis/2010/08/12/tasteful-in-times-square-theres-a-law-against-that/

Steiner, F. R., & Butler, K. (Eds.). (2007). *Planning and urban design standards, student edition*. Hoboken, NJ: John Wiley & Sons.

Testa, B. (2016, February). Sign of the times. *Planning*, 19–23, 22.

Thompson, A. (2012). Signs and billboards: What's legal and what's not? *CityLand*. Retrieved from http://www.citylandnyc.org/signs-and-billboards-whats-legal-and-whats-not/

Weinstein, A. (2001). Legal issues in the regulation of on-premise signs. In M. Morris, M. L. Hishaw, D. Mace, & A. Weinstein (Eds.), *Context-sensitive signage design*. American Planning Association Planning Advisory Service. Retrieved from https://planning-org-uploaded-media.s3.amazonaws.com/legacy_resources/research/signs/pdf/contextsensitivesignagedesign.pdf

Weinstein, A. C., & Connolly, B. J. (2015). Sign regulation after reed: Suggestions for coping with legal uncertainty. *The Urban Lawyer, 2015*. Cleveland-Marshall Legal Studies Paper No. 15–285. Accepted paper. Retrieved from http:ssrn.com/abstract=2660404

PART TWO

Place and Regulation

CHAPTER FIVE

Public Space and Communication

The Zoning of Public Interaction—Revisited 2016

SUSAN J. DRUCKER
Hofstra University

GARY GUMPERT
Urban Communication Foundation

Introduction

We have marked 100 years since New York City passed one of the nation's first Zoning laws on July 25, 1916. It was the first set of land-use rules in the U.S. applicable to an entire city. Over the years zoning laws have been attributed with being valuable public health and safety initiatives and a form of social engineering. The Museum of the City of New York's 2016 focuses on the consequences and the importance the first U.S. zoning law:

> The character of New York's varied neighborhoods is governed by a novel set of rules first envisioned by New York reformers 100 years ago—the groundbreaking Zoning Resolution of 1916. Zoning, which was designed to tame the unruly process of free-market real estate development, has continued to shape the city we know today in countless, often unseen, ways.
>
> This landmark law gave birth to the iconic "setback" skyscraper and the modern skyline; to special neighborhoods like the Theater District; to public amenities like pedestrian plazas, and to residential neighborhoods of all shapes and sizes. On the 100th anniversary of America's first comprehensive zoning resolution, Mastering the Metropolis: New York and Zoning, 1916–2016 will examine the effects of the evolving law and chart the history of the city's zoning rules and debates to the current day, illuminating

> how the tools of zoning have reflected a century of evolving ideas about what constitutes an "ideal" city. (http://www.mcny.org/exhibition/mastering-metropolis)

That document has been praised for creating a planning system as well as a zoning law based on "accepted common-law and constitutional doctrines: that landowners are not entirely free to create nuisances to those around them; and that local governments may police conduct in the name of public health, safety and welfare" (Dunlap, 2016). It has been created with reducing urban congestion and preserving light and air. It has been vilified for increasing income segregation (Fox, 2016) and segregation from "the other," at that time the growing immigrant population was "deemed 'undesirable'" (Graham, 2016).

Retrospectives on zoning brought on by the centennial anniversary have been many, reflecting on the "profound effect" of the zoning resolution on the built environment and changes in zoning over the years (Weisbrod, 2016). Carl Weisbrod, Director of the New York City Department of City Planning and Chair of the New York City Planning Commission observed:

> … with the 1961 Zoning Resolution, zoning underwent a fundamental change. It went from being a control over what should not happen, to being a positive document. In simplest terms, it was a reflection of its time and channeled what should or could be built in a district. It sought to inject open spaces—and parking spaces—into the body of a city that was expected to redevelop rather than grow. (Weisbrod, 2016)

Critics have noted the tensions of growth and management (Weisbrod, 2016), the resulting effect shaping the city skyline (Eide, 2016) and the creation of green spaces as a result of zoning (Zoning NYC Scavenger Hunt, 2016). The creative interpretation of zoning laws have been said to actually encourage mixed-use zoning rather than segregated uses.

Back in 1991 the authors argued that zoning laws were powerful determinants of urban communication since regulations shaping the physical environment have profound interactional consequences. We return to the position that zoning is a major tool of planning goals and development. Long neglected is the tremendous impact on communication these regulations have had as they shaped the built environment and activities within that environment. Zoning laws illustrate the powerful symbiotic relationship between contexts of communication and regulations which directly and indirectly govern communicative spaces. The urban landscape has been changed, perhaps revitalized by the connected, mobile media environment. An assortment of technologies enable information from the digital world to be layered onto the physical world altering the person/environment relationship by creating spaces in which users interact with their physical surroundings through digital media. The line between physical and virtual environments is blurring. In the intervening years between our original thoughts on zoning and our return to the subject we have argued that there is a reciprocal relationship between physical and media space in which one is layered upon the other (Gumpert & Drucker, 2010). Mobile devices have ushered media consumption into public spaces, augmented by ubiquitous/pervasive mobile computing and wireless connectivity. Mobile media of old from books and newspapers to transistor radios and walkmen have given way to smart phones, handheld games, laptops, tablets, watches and internet connected fitness trackers. The person/environment relationship has been shaped by and reflects the digital world but our observations about the power of zoning in shaping the urban communication environment remain valid. We share those long ago thoughts with you.

Public Space and Communication: The Zoning of Public Interaction—1991

Although interaction has been emancipated from place, public places still function as sites of face-to-face interaction. Today, spacial propinquity, the actual physical nearness of individuals, is less important to the maintenance of social communities than in the past because the modern media have eliminated "cohabitation of a territorial place" as a prerequisite for community (Gumpert, 1987, p. 78). With few exceptions, the adult urbanite or suburbanite is increasingly able to maintain selected contacts with others on an interest basis over greater distances and is thus a member of an increasing number of communities that are not territorially defined (Webber, 1964). Such associations are called "interest communities," "nonplace communities," "nonpropinquitous communities," or "media communities" (Gumpert, 1987). In short, relationships are no longer restricted to place but are distributed through space.

Changing communication relationships accompany the changing nature of American towns and cities. The dynamically changing population base, the shift from an agrarian economy to an industrial one, and the transportation revolution of train, car, and plane have altered both psychological and physical attitudes toward time and space. However, the presence of a communication revolution is seldom mentioned as a stimulus of urban social change, and yet the acceleration of media development that serves to emancipate people from the tyranny of place has certainly contributed to the present state of relations between person and place.

The media offer and Americans wholeheartedly accept alternative forms of interaction. In a nation where approximately 99% of all homes have radio and television sets, (70% with videocassette recorders; *The broadcasting yearbook*, 1990), and 93% of all households have telephones (31% with answering machines; Federal

Communications Commission, 1990), the alternatives to face-to-face communication have shifted much social interaction into the mediated realm. However, mediated social interaction and face-to-face interaction, while functionally equivalent, are contextually and experientially dissimilar.

Face-to-face communication offers full sensory involvement, while mediated communication only uses, by definition, one or more of the senses, never the complete sensorium. Thus, while specific communicative tasks may be success fully completed, the degree of sensory involvement will be limited. In addition, mediated interpersonal communication is generally task oriented, thereby limiting the degree to which unpredictable interaction might occur since circumstances are circumscribed. While in many cases electronic communication may be more efficient and convenient than face-to-face communication, the members of a well-adjusted society retain interactional possibilities in which face-to-face social interaction in public remains one of the options. The public environment provides emotional prospects not available through media technology in the privacy of one's home. According to the environmental psychologist Albert Mehrabian, environments can generate feelings of anger, fear, boredom, or pleasure (Mehrabian, 1976, p. 10). The subtitle of sociologist Ray Oldenburg's The Great Good Place is particularly significant: Cafes, Coffee Shops, Community Centers, Beauty Parlors, General Stores, Bars, Hangouts, and How They Get You Through the Day (Oldenburg, 1989). The message is clearly that people need public as well as private space.

> In the absence of an informal public life, Americans are denied the means of relieving stress that serve other cultures so effectively. We seem not to realize that the means of relieving stress can just as easily be built into an urban environment as those features which produce stress. (Oldenburg, 1989, p. 10)

A recent study indicates that public places sustain therapeutic environments. Results of a pilot study conducted by environmental design professors Carolyn Francis and Clare Cooper (1991) provide evidence that people need places to which they can take their problems, including locales with viewpoints, views of nature, and places to be alone in public. Michael Brill (1989) argues that "we do not recognize much of our public life that really is here, nor do we recognize how much of it has moved into 'virtual' space, the space of electronic media" (p. 17).

Mediated communication is qualitatively different from face-to-face interaction. Mediated functional alternatives generally occur inside controlled private space. Richard Sennett, in his (1990) The Conscience of the Eye, describes the creation of "bland neutralizing spaces, spaces which remove the threat of social contact . . ." (p. xii). Media technology facilitates the transformation of social interaction from the public to the private domain.

Public and Electronic Space

Not so long ago a community was a geographic arena in which the major social activities of commerce, production, and residence took place. Changes in social activity and communication patterns accompanied urbanization, automobiles, and advances in electronic media. "The technology of modem motion replaces being in the street with a desire to erase the constraints of geography" (Sennett, 1987, p. 32). Workers and employers who had lived and worked close to where goods were manufactured or services supplied were freed from the need for such physical presence. Functionalism, environmental planning based on function, segregated cities by uses to "assure the individual healthy living conditions and to distribute the physical benefits" (Gehl, 1987, p. 45). Yet the functionalists did not consider the social implications of design and planning.

> Throughout the entire history of human habitation, streets and squares had formed focal points and gathering places, but with the advent of funtionalism, streets and squares were literally declared unwanted. (Gehl, 1987, p. 47)

Commerce, production, and residence each elevated the, importance of either privacy or interaction. Interaction became paramount in commerce while privacy became all important in residence. The eventual division of these into separate activities resulted in the creation of (1) a sphere of privacy and (2) a sphere of interaction.

As the work-residence-commerce configuration changes, the opportunity for accidental and incidental interaction with others is also altered, but not the need for such interaction. As that configuration changes, interaction becomes increasingly programmed and functional. Rare are the possibilities of unplanned interludes of people watching and potential relationships. Yet, where people gather in public, where they congregate and drink, where they go to watch others, is the province and direct and indirect result of legislative and administrative land-use decisions. "As such, these laws and regulations actually do more than determine how land is to be used and arranged, they actually structure the ways in which people, as social groups, relate to each other" (Schneekloth, 1987, p. 318).[1]

A number of theorists and critics have commented on the relationship of urban planning and social interaction (Appleyard, 1981; Brill, 1989; Dovey, 1985; Francis, 1987; Gehl, 1987; Jacobs, 1961). In his analysis of the medieval town Louis Mumford (1987) illustrates how the walls protecting the town also provided an "open promenade for recreation" (p. 66) and that "the street was a line of communication for pedestrians" (p. 69). J. B. Jackson (1987b) explains that the reason that the street has not been widely regarded as a medium of communication is because today "we usually experience it by moving through an elaborate system of arteries that has priority over the environment of building" (p. 76).

In her pioneering work The Death and Life of Great American Cities (1961), Jane Jacobs recognizes the critical need for social interaction but fails to connect the state of the urban environment with an additional source of influence-the acceleration of media technology.

> When an area of a city lacks a sidewalk life, the people of the place must enlarge their private lives if they are to have anything approaching equivalent contact with their neighbors. They must settle for some form of "togetherness," in which more is shared with one another than in the life of the sidewalk, or else they must settle for lack of contact. Inevitably the outcome is one or the other; it has to be; and either has distressing results. (pp. 62–63)

Similarly, while Bibb Latané (1981) proposes a theory of social impact based on the effect of other persons on an individual, that theory fails to recognize media technology as a social source and variable that may determine the presence or absence of others.

In the area of ecological psychology Roger Barker (1968) distinguishes be tween the "psychological environment ... the world as a particular person perceives and is otherwise affected by it" and the "ecological environment ... the objective, preperceptual context of behavior; the real-life settings within which people behave" (p. 1). What Barker does not recognize is the dynamic nature of the psychological environment, which is altered and shaped by the allocation of media resources. The relationship between the psychological and ecological environment is not only complex and symbiotic but reciprocal.

Many other writers have documented and described the street as a place of interaction, transaction, and communication (Czarnowski, 1978; Eichner & Tobey, 1987; Jackson, 1987b; Rudofsky, 1969; Vidler, 1978). A host of other critics have shown an understanding of the intimate consequences planning and architecture have upon social interaction (Clay, 1987; Rykwert, 1978; Sennett, 1974; Whyte, 1988), but few, other than Michael Brill (1989), recognize the links between planning, social interaction, and the development of media technology.

The influence of architecture and design on social interaction has been explored, as have the effects of media technology upon the quality of and opportunities for communication (Gumpert, 1987; Manheim, 1987; McCombs & Shaw, 1972; Pool, 1977; Shaw & McCombs, 1977).[2] Research has centered on issues of space and interaction, the regulation of media technology, and ways in which environmental attributes communicate and guide behavior (Burgoon, Buller, & Woodall, 1989; Goffman, 1963; Hall, 1966). The communication literature has examined the impact of media alternatives upon face-to-face communication (Altheide & Snow, 1979; Carey, 1988; Klapper, 1960; McQuail, 1983; Schramm,

1973).[3] While the communication discipline has begun to examine the effects of media on social interaction, it has neglected the resulting changes in patterns of face-to-face interaction. Architecture and design, media technologies, and patterns of social interaction influence each other in complex ways. They coalesce when government regulates design and construction and thereby social interaction. The primary mechanism for the implementation of government land-use policy is through zoning laws.

The History and Purposes of Zoning

Zoning laws regulate face-to-face interaction restricting or stimulating communicative activity either explicitly or implicitly. "Land use planning, zoning, and development practices are a shorthand of the unstated rules governing what are widely regarded as correct social categories and relationships" (Perin, 1977, p. 3). Zoning laws that partition cities into areas based on function are predicated on the existence of both transportation and media systems that provide movement and communication between precincts. Zoning is the legislative tool that limits the use of property (Mandelker & Cunningham, 1985). Public controls on the private use of land can be traced to the beginnings of civilization (Bescher, Wright, & Gitelman, 1976). Aristotle's description of the ideal public square in The Politics suggests zoned activities. He describes the agora or "free square" as a place for discussion and exchange of ideas but also as a site from which all commercial activities and all merchants and vendors should be exiled (Aristotle, 1988, p. 425). "[I]t is here in the agora or forum that history is made visible and where speech becomes a political instrument, eloquence a form of political action" (Jackson, 1987a, p. 120). Merchants were generally excluded from the forum in classical cities as this was a space reserved for "a superior order of citizens: for political action and the exchange of ideas" (Jackson, 1987a, p. 120).

Concern for health and safety led to laws creating use districts within a city. Ancient Rome recognized the need for zoning of areas and the regulation of building height within those areas (Yokley, 1978).[4] In England, this approach to land-use regulation dates back to 1581 in an act that limited "making of iron metal in iron-mills within the compass and precinct of two and twenty miles from the city of London or its suburbs" (Bescher et al., 1976, p. 5).[5]

Authorities differ as to the origins of modem zoning.[6] A German authority, Baumeister, published a treatise in 1876 reviewing the early history of zoning and records "use" zoning as having been founded by a decree of Napoleon I in 1810. Napoleon's order sought to protect districts from the invasion of injurious uses (Metzenbaum, 1955).[7]

In the United States urban planning was initiated by the first settlers who conceived of planning as laying out streets, locating public buildings, and implementing systems for the distribution of land.[8] A 1692 enactment by the Province of Massachusetts Bay regulated nuisances created by slaughterhouses, still houses, tallow chandlers, and curriers. In the same year Boston established fire zones and many cities followed suit (Metzenbaum, 1955).[9]

Regulations and community planning reflected concern for health and social conditions. Population density was regulated through "tenement house codes." Building codes, sanitary codes, height ordinances, and nuisance ordinances followed as cities came to perceive a need to protect the public welfare (Metzenbaum, 1955).[10]

New York City is credited with the first modem zoning law in the United States, significant as the first complete and comprehensive system of building control and a model for the entire nation.[11] It established districts throughout the five boroughs and restrictions on uses permitted for property within the city, building height, and the proportion of area available for building (Bescher et al., 1976). It provided for three separate classes of use districts: (1) residence districts, (2) business districts, and (3) unrestricted districts. Residence districts prohibited all trade and industry while business districts only excluded specified "nuisance" types of businesses such as manufacturing of ammonia and paint (Bescher et al., 1976, p. 502).

"By 1922 it was reported that twenty state zoning enabling acts and fifty municipal zoning ordinances were in force or in process of formulation" (Mandelker & Cunningham, 1985, pp. 143–144). By 1926 the Supreme Court decision in Euclid v. Ambler Realty Co. (1926) firmly established zoning by districts or the division of the community into districts. Land uses were then determined on an "as of-right" basis. "As of-right" meant that after the formulation of the ordinance text and map by a local zoning commission and its adoption by the local governing body, most administrations would require only the services of a building official who would determine whether proposed construction complied with the requirements.

The objectives of zoning have included protection of property values, preservation of the character of neighborhood, insurance of adequate governmental services, minimization of traffic congestion, aesthetics, and historic preservation. Zoning divides the municipality into districts and prescribes requirements within each district dealing with use (i.e., industrial, commercial, residential), density (i.e., number of units per acre), and siting of development on each parcel (e.g., setback or landscaping requirements).

Greater discretion developed within the administration of zoning ordinances as a result of post-World-War-II urban sprawl, which forced planners to seek techniques with which to improve the quality of city life. "They were asked to enact land use controls which would create and preserve a healthy, aesthetically attractive community with proper regard for the finite quality of resources and the historical patrimony of the community" (Sweet, 1987, p. 206). This led to a more flexible policy encompassing variances that permitted a parcel of land to be used differently than prescribed by zoning ordinances as well as the increased issuance of special-use permits (Sweet, 1987).[12]

Zoning and Interaction

No law operates in isolation; zoning laws were and are no exception. Every zoning law has consequences for social interaction. Zoning regulations influence communication by controlling communication activities or by controlling communication contexts.

The recognized way in which zoning has been linked to communication activities involves the direct exclusion of certain communication activities from some districts.[13] Such laws determine the lawful location of theaters, the clustering or dispersing of adult bookstores, or the placement of billboards and signs.[14] While the question of communication activities has been dealt with extensively by scholars of communication law, little analysis has been done on the impact of zoning, which, while regulating land use, regulates communication contexts.

Social interaction, zoning, and media development are interrelated and inter twined. Thus, while zoning in the name of public welfare and safety regulates and controls social interaction, media development facilitates communication and affects the sites of interaction. Both zoning and media development influence the context and qualitative nature of social interaction and of each other.

Zoning regulations influence communication by controlling the physical con text of communication. Several provisions of the zoning law of New York City and its suburbs demonstrate ways in which zoning implicitly restrains or stimulates communication. Two symbolic examples serve to illustrate: setbacks, the physical space between private structure and public thoroughfare, and sidewalk cafes, sites with which communication has long been associated.

Protection from Interaction: The Setback

How far should dwellings and businesses be located from the street? Does depth of the space influence social behavior? Most zoning ordinances specify that new buildings be constructed within a setback line. The setback space separates the activities of one sphere from another, building from sidewalk or street. While setbacks always exist, zoning laws mandate the particular dimensions to protect commercial vistas and provide space for front yards in residential districts (i.e., aesthetic objectives) (New York City Department of City Planning 1988, article 3,

§7.10). The depth of setback and width of sidewalk either encourages or discourages interaction.

In New York City residential setbacks range from 20 feet (Residence District 1) to 15 feet (Residence District 2 and Residence District 3).[15] In suburban residential areas setbacks reflect or set the agenda of a community that prizes privacy over connection. Towns and villages have zoned minimum setback requirements ranging from 15 to 50 feet (Town of Oyster Bay, 1989, §217; Village of Great Neck, §210–50 and 210–63; Village of Great Neck Plaza, 1989, article 3). A residential structure set at these distances from the front line of the building to the front line of the street or next building produces a buffer strip that helps reduce traffic noise and creates a private environment deemphasizing the presence of neighbors. The setbacks offer a vantage point for the "safe" observation of the activities of the street. The depth of setback establishes a communication context or social agenda that can either deter or facilitate communication with others.

Some architects assert that the closer one lives to the street the more one is likely to participate in street life. The distance of a home from the street becomes a neutral zone in which social interaction seldom occurs and is more probably discouraged (Moore & Sullivan, 1988). It is likely that the depth of that zone is a variable that can either facilitate or deter interaction. For the International School architects, visual connection was linked more to detachment than interaction. These architects were dedicated to the concept of visibility in construction in which a wall of glass isolates the activities within the building from the life of the street. The aesthetics of visibility and social isolation merged. Setbacks allow control of privacy, allowing for visual interest shielded by social isolation. In The Fall of Public Man, Richard Sennett (1974) asserts,

> Human beings need to have some distance from intimate observation by others in order to feel sociable. Increase intimate contact and you decrease sociability. … Isolation in the midst of public visibility and overemphasis on psychological transactions complement each other. (p. 15)

In order to be social, some facade of the private must be preserved. The public space requires the possibility of privacy in order to enhance sociability. The setback provides a paradoxical degree of control: to insulate from or stimulate interaction.

Setbacks may actually encourage social interaction—if coupled with complementary design characteristics and if the ideal distance is maintained between structure and site of potential interaction. The amount of setback that might encourage interaction is determined, in part, by the psychological and social attitude generated by proximity to the street. Does the street offer visually inviting, entertaining, nonthreatening activity? If so, then a suitable setback would be that distance from structure to street that offers a clear view while offering a sense of detachment.

The premise of the setback is to establish a buffer, a neutral zone, separating the activities of the street from the private affairs of home and office. The intent is not to sever connection but rather to control the impact of public upon private functions through separation. The implicit assumption is that physical space is a variable that if controlled would protect and separate the inside from the outside, the private from the public.

The windows and balconies of high-rise apartment buildings are the urban equivalents of suburban setbacks. They provide inhabitants the luxury of an aesthetically distanced and sometimes voyeuristic relationship with the outside world. But height has its limits and beyond four stories potential interaction is sacrificed for sunlight and panorama (Alexander, Iskikawa & Silverstein, 1977).

The Electronic Setback

While the explicit assumption behind media development is connection not separation, the developmental and implicit consequence is similar to that of the setback; but instead of creating a buffer zone, an electronic wall is constructed that thereby defines the exterior as inconsequential. Media of communication create psychological "nonplace" setbacks that are in creased and decreased at will. With increasing dependence of individuals upon media for news, information, entertainment, and interpersonal interaction, the architectural dimensions of setback may become insignificant. If the psychological focus is through the media of the home, from the inside, the physical distance between the public and the private can become irrelevant.

Facilitating Interaction: The Cafe as Medium

There are a variety of places in the United States where some form of public social interaction occurs. Bars, taverns, coffee shops, diners, and community centers represent some of the obvious sites for public interaction while beauty parlors, barbershops, laundromats, shops, and libraries comprise locations for less planned contacts. Of these, bars, coffee shops, and diners serve food and drink and provide limited opportunity for interpersonal contact and conversation due to the economics of place. Steady consumption while on the premises, rapid turnover of clientele, and increased dependence upon food as a source of revenue are practical necessities for the modern American bar or tavern. Throughout the rest of the world cafes serve a variety of social functions deemphasized in the bars, taverns, and coffee shops of the United States.

The cafe is a medium where business is conducted and news is exchanged, and where there is an opportunity to share reactions to events. It is a meeting place, a place in which to be entertained (i.e., with chess, backgammon, cards), a place to avoid loneliness, a place to watch television or listen to the radio with others, a place to read newspapers, a place to write, a place to belong, a place with which one may be identified, and a place to which one can escape. In Europe the cafe has long been connected with the medium of the newspaper. To this day, the typical Viennese cafe provides approximately 20 different newspapers for its patrons. Not only is it the center for information conveyed interpersonally, but it serves as a library and reading room. In the United States economics have, for the most part, altered the traditional cafe. Cafes relegated to high rent commercial and manufacturing districts, locations associated with work rather than leisure, stress dining and transaction. While public leisurely reading is deemphasized, the cafe (and its many variations) still provides the opportunity to commune with others.

Of all of the potential sites for interaction in New York, cafes (traditional storefront, enclosed outdoor, and open-air sidewalk) are specifically singled out by zoning laws. Thus, traditional cafes are prohibited from any and all residentially zoned areas. The Zoning Resolution of the Department of Planning of New York City (1988, Chapter 4) addresses the regulation of both enclosed and sidewalk cafes. Its general purpose is "to establish city-wide regulations designed to encourage sidewalk cafes in locations where they are appropriate and discourage them in locations where they are inappropriate and to promote and protect public health, safety, general welfare and amenity. … Sidewalk cafes may be established only at the discretion of the City of New York pursuant to §14–02" (N.Y.C. Department of City Planning, 1988, §14–02). The cafes of Paris, the pubs of England, the coffeehouses of Vienna, the kafenia of Greece are not restricted to commercial areas and function as meeting places, sites of interaction, and perches from which leisurely to observe neighbors. Yet in New York City "no sidewalk cafe shall be permitted in residential districts" (N.Y.C. Department of City Planning (1988, §14–04). Such potential functions of the local cafe have been "zoned out" by New York City regulations.

However, "sidewalk cafes shall be permitted in street malls" and "may be permitted in Historic Districts or in designated Landmark Buildings" (N.Y.C. Department of City Planning, 1988, §14–04). As a result these sites of interaction have been relegated to "pseudo" streets, to becoming artificial oases without permanent constituency. But no sidewalk cafes or enclosed cafes are permitted on most major thoroughfares in Manhattan (N.Y.C. Department of City Planning, 1988, appendix A and B).[16] In New York City although sidewalk cafes may be permitted in all commercial and manufacturing districts, cafes in certain areas are nevertheless still prohibited (i.e., the theater district, Lincoln Square district,

United Nations area, Fifth Avenue, and Madison Avenue) for the stated purposes of insuring adequate space for pedestrians on the sidewalk, preserving and enhancing the character of neighborhoods, and promoting the most desirable use of land and conservation of the value of land and building, thereby protecting the City's tax revenues (N.Y.C. Department of City Planning, 1988, §14–00).

Suburban cafes are governed by the same regulations that apply to restaurants. In Great Neck, New York, any shop or store that makes available any facility for, or which permits the consumption of, food or beverages on the premises outside a building is prohibited, as are open-front shops and stores (Village of Great Neck, 1988, §210–102 (C) & 93A). In some central business districts services are permitted on an adjacent terrace (Town of Oyster Bay, 1989, p. 1424). Permitted occupancy of restaurants is specifically defined based on size, type of seating, and activities (e.g., "places containing permanent seats, 1 person per 6 square feet, designated waiting areas, not including exits and hallways, 3 square feet per person, counter dining, 1 person per counter stool or seat at 1 stool or seat per 3 foot length of counter space" (Town of Oyster Bay, 1989, p. 1408)).

The restrictions shape not only the physical but the psychological map of the locations. The inherent and controlling view manifested in zoning regulations has become that of the street as conduit rather than destination, a place in which unimpeded traffic (both pedestrian and vehicular) must be preserved. Sidewalk cafes by definition extend into the street and a series of regulations imposed upon activities in the street affect permissible behavior within the cafe. In New York City the "regulation of cafe boundary" mandates that "no portion of side walk cafes, such as doors, windows, walls or any objects within the cafe may swing or project beyond a designated exterior perimeter" (N.Y.C. Department of City Planning, 1988, §14–21). Clearly the intent is to prevent the structure from impeding traffic.

The acoustical ambiance of the cafe does not deter the flow of pedestrians, yet sound is regulated by "unnecessary noise" restrictions that prohibit the use of "musical instruments or sound reproduction devices" within a sidewalk cafe for any purpose (N.Y.C. Department of City Planning, 1988, §14–26). When the environment of the cafe expands to include a portion of the sidewalk, the govern ing regulations stress the sidewalk nature of the establishment. Thus, zoning is imposed to regulate interior communication contexts. The potential intrusion of sounds emanating from the cafe justifies the regulation of the sound environment of the cafe.

The notion of seclusion within visibility (similar to the general precept of the International School regarding walls of glass) extends to the cafe and its integral functions and relationship to the street. The cafe functions as a place to watch and be watched. The literature of research in environmental design has established that windows provide four psychological benefits: (1) access to environ mental information such as time, weather, and seasonal changes; (2) access to sensory change; (3) connection to the world outside, access to events beyond the walled boundaries; and (4) restoration and recovery, release of tension (Heerwagen, 1990). All of these communication-related functions are incidentally facilitated by zoning regulation. The New York City Zoning Resolution (1988, §14–32) requires that in cafe's all enclosing walls, doors, and above floor level or base walls and must be colorless untinted, nonreflective transparent material.[17] "At least 50% of the walls, up to a height of 7′0″ above finished floor level, shall consist of operable transparent" (N.Y.C. Department of City Planning,1988, §14–32). This regulation encourages the very kind of correlation and surveillance of the environment that enhances the communicative functions of the cafe.

The cafe is an institution and medium integrated into the fabric of social life worldwide. It is a natural part of the public landscape fostering normal, com mon, familiar services and contacts. But in the United States it is sequestered and removed from the everyday and extracted from the routine. The cafe and its inherent social functions are specifically controlled by land-use regulations.

Segregating Uses and Interaction

In A Pattern Language, Alexander et al. (1977) state that "the artificial separation of houses and work creates intolerable rifts in people's inner lives (p. 52). The zoning of "residential use districts" affects the density of population and cost and availability of housing, thereby circumscribing the income level of in habitants. Legal decisions do not endorse such an exclusionary purpose,[18] but the de facto result of such restrictions affects the make-up of a community's population (N.Y.C. Department of City Planning, 1988, chapter 3, §7.24) and its quality of life. Restaurants, schools, and places of worship are typical locations of interaction and special permits are required to enable such establishments to exist (N.Y.C. Department of City Planning, 1988, chapter 3, §8.17). Even in a commercial area, zoning ordinances delineate the permissible types of businesses. For example, in New York City there are eight different types of commercial districts, each providing for specific types of commercial activities (e.g., local retail districts, local service districts, general commercial districts, general and restricted central commercial districts) (N.Y.C. Department of City Planning, 1988, article 3). The districts are designed to promote commercial development while preventing congestion, health and safety hazards, and unwanted changes in the character of certain areas.

Some urban planning commissions, aware of the need to develop places for social interaction, have used incentive zoning to create sites of interaction. Real estate developers have been economically encouraged to create open spaces and urban plazas for public use.[19] The builders have been permitted to build over zoning limits if they provide a public plaza, or an arcade, or a comparable amenity. In New York City between 1961 and 1973 some 1.1 million square feet of new open space was created through the use of incentive zoning (Whyte, 1988).[20]

Although a step in the right direction, the results of incentive zoning have been problematic as the public places created are located within areas still segregated by use due to underlying zoning. Thus, "bonus plazas" (the name given the results of incentive zoning) dedicated to interaction serve a commuting population and are left deserted after work hours.

The Mediation of Public Space

Public places, illustrated by the spaces created by setbacks and cafes, provide sites not only for public interaction but for media use. Increasing individual dependence upon media technology for the purposes of information, entertainment, and communication affects both setbacks and cafes, because less reliance and importance is placed upon the communication functions of public space. However, the impression that public social interaction and media usage are antithetical is misleading and inaccurate. Media use may be incorporated into public space in three ways: (1) further to privatize public space (e.g., through the use of a Walkman, Watchman, or cellular phone); (2) to impose or demonstrate control of space (e.g., with large portable stereophonic systems); and (3) to entertain and inform the public by the incorporation of media technology into private and public establishments (e.g., through the use of television in bars and pubs, and of giant video boards such as are found in New York's Times Square).

Electronic space has been made portable and personal. "Portable public privacy ... [is] an expression of a particular vision of the world" (Gumpert, 1987, p. 87). Public media consumption has followed a pattern of privatized use. Persons walk through the streets or publicly sip a cup of coffee connected to an acoustical world of their choosing outside yet insulated. While portable stereo systems (also known as ghetto blasters or boom boxes) may express some degree of territoriality, they are perceived by some as aggressive and have been discouraged by police action based on noise control codes and loitering laws (McKinney's Consolidated Laws of New York Annotated, 1987). Paradoxically, media technology has been integrated into the activities of the public sphere, although it has been one of the forces that deemphasize the demand and need for public social interaction. While zoning laws have discouraged or depopulated public spaces, public use of electronic media has further diminished the social vitality of those places. The challenge for designers is to discover ways of introducing "public media" in the public spaces available to a population weaned on media.

Conclusion

Public interaction is at times a direct factor in planning but is frequently viewed as an incidental consideration, and in some situations it may be merely an un planned outcome. Each act of planning and each architectural design has an interactional impact upon individual and society; each communication technology has a similar interactional impact. Communication technology, planning and design are symbiotically linked influencing one another in enigmatic, paradoxical, and complicated ways. Design and planning (enacted through zoning) affect social interaction as electronic media of communication alter the human demands made upon the physical ecological environment. Zoning is not merely a mechanism for control of land use, it is implicitly connected to social networks.

There is a reciprocal relationship between zoning and the media of communication. The zoning of public interaction is predicated upon a model of social interaction in which social impact is shaped by this interrelationship. Contemporary public interaction is influenced by media development. Modern zoning and regulations that govern public space affect public social interaction. Media use and developments combine with zoning to recast the profile of contemporary public interaction.

Originally, zoning laws were formulated to protect the public welfare, a term long associated with the physical health and safety of a community. The concept of public welfare should be enlarged to encompass the psychological health of the community in light of the formidable impact of zoning on interaction. There should be explicit recognition that zoning laws must be examined as significant in controlling the opportunity for interpersonal communication. Presently, zoning laws regulate both communication activities and the context of communication yet reflect an underestimation of the continued importance of face-to-face interaction.

Media affect perception of place and use of space. Planners and land-use regulators have joined media theorists by failing to integrate the impact of media on face-to-face interaction in urban and suburban environments. The challenge is to choreograph both public space and the newer forms of electronic space.

There is often a gap between academic research and practice. But in the establishment of a paradigm of public interaction in which interpersonal and mediated communication are but two of a number of variables, the gap becomes a chasm. If the nature of public social interaction has receded into the realm of privacy, and such a phenomenon is not in the best interest of human beings, then the challenge for communication scholars should be to recommend and apply communication principles that will help change the status quo.

What are the communication principles that can be applied to planning, design, and regulation of the ecological environment? How may one forecast the communication implications of design, planning, and zoning? What are the implications of media use for design and planning? More specifically, is there a relationship between private mediated interaction and safety in public places? Is there a relationship between the distance from the street (vertical or horizontal) and media consumption? Do different cultural patterns emerge with regard to media use and penetration and communicative functions of public places? Would the integration of media technologies into public spaces change the relationship to, and use of, such spaces? Can zoning be used to create "mixed-communication-use zones" simultaneously making available public mediated and face-toface interaction?

Although until recently the majority of the world's population inhabited villages, it has been predicted that by the end of this decade the majority will live in cities and within 100 years, 90% of the worldwide population will live in cities (Hiss, 1990). The public experience of the city should be designed to enhance social interaction.

Notes

1. The scope of this article is limited to zoning although other land-use regulations such as nuisance laws and restrictive covenants also have an impact upon public interaction.
2. Joshua Meyrowitz in No Sense of Place notes that changes in the means of communica tion contribute to social change but that such influence is often overlooked (Meyrowitz, 1985). Meyrowitz provides one of the few published inquiries into the ways in which electronic media have affected social behavior through the reorganization of social settings and the alteration of the relationship between physical place and social "place."
3. The majority of media studies have emphasized the technology, the process, and the broad social effects rather than investigating the much more complex interrelationship of technology, social forces, and interpersonal interaction.
4. The earliest code of Roman law, the Twelve Tables (drafted in 451–445 B.C.) provided, Whoever sets a hedge around his land shall not exceed the boundary; in the case of a wall, he shall leave one foot; in the case of a house, two feet. If a grave or pit, the required depth. If a well, a path, an olive or fig tree, nine feet. Finally, whoever plants other trees shall leave a space of five feet between [his] property and his neighbor's. If there is litigation about boundaries, five feet (Bescher et al., 1976, p. 2). This was enacted by the "Act touching Iron-Mills near unto the City of London and the River of Thames" (Bescher et al., 1976, p. 5).
5. Thirty-five years later the Prussian Code provided a more comprehensive development of the zoning concept that was developed by laws of the German Empire. These restrictions on the use of land gained popularity and were adopted in principal cities such as Frankfurt,

Berlin, and Coblenz. According to Harold M. Lewis in City Planning, others trace zoning as beginning in Germany in 1884 (cited in Metzenbaum, 1955).

6. In England land-use restrictions in the modern era were initiated through the official action of the "Town Planning Act" of 1909, which combined zoning and planning. In 1919 Parliament passed an act that made town planning and regulatory legislation, previously discretionary, mandatory after 1923 (Metzenbaum, 1955). See Part 2 of 9 Edward VII, chapter 44 and Part 2 of 9 and 10 George V, chapter 35 of the Town Planning Acts of 1909 and 1919, Law Reports Public General Statues of 1919 (cited in Metzenbaum, 1955).
7. Planning in the United States reflected the European orientation of planning. Settlements in the southwest were based upon traditional Spanish planning of a gridiron with a central plaza while Louisiana was patterned after French design of string villages divided into long lots perpendicular to the Mississippi River (Anderson, 1986, p. 10).
8. In colonial New England where Puritanism emphasized community and order, virtually all early towns centered around a compact core of dwellings, a fort, and a public meeting house. The General Court fixed the "shape and size of the own" (Hawke, 1966). Some early New England settlements (such as Dedham, Massachusetts) used land grants as a means to "shape as well as to limit the ideal society" that founders sought to create (Lockridge, 1970).
9. These restrictions were often recognized as within the inherent police powers of municipalities (Bassett, 1936). The United States Supreme Court upheld a Massachusetts "fire control" measure (passed in 1904–1905) that fixed height limits in certain residential areas in the city of Boston (Welch v. Swasey, 1909). In 1909 California enacted a system of zoning under authority of its state constitution. The City of Los Angeles passed an ordinance dividing the city into industrial and residential districts prohibiting certain buildings and uses such as laundries (Ex parte Quong Wo, 1911) and brickmaking within a designated area (Hadacheck v. Sebastian, 1915).
10. The terms zoning and planning are not synonymous. Zoning is a separation of the municipality into districts and the regulation of buildings and structures in the districts in accordance with the nature of their use. It pertains to the use as well as the structural and architectural design of buildings. Planning may be defined as a broader term connoting a systematic development designed to promote the common interest of a community such as streets, parks, industrial and commercial undertakings, civic beauty, and other matters within the police power (Yokley, 1978).
11. A variance is permission to use or build on land in a manner contrary to applicable zoning regulations. Two types of variances are use variances (those that allow a type of activity prohibited by zoning law) and area variances (those that allow a change in dimensions required such as a setback) (Willis, 1989). Variances may be contrasted with special permits, which involve a permitted use under a zoning law but the use is only permitted when an applicant has satisfied standards established in the zoning law (Sharp, 1989).
12. Communication scholars have generally addressed the relationship of zoning and communicative activities in the context of communication regulatory issues and freedom of speech case law (Gillmor, Barron, Simon, & Terry, 1990; Middleton & Chamberlin, 1988; Pember, 1990). For an introduction to the considerable literature on zoning and the regulation of communication activities see Sign Regulation for Small and Midsize Communities (Kelly & Raso, 1989); Street Graphics and the Law (Mandelker & Ewald, 1988); City

Signs and Lights: A Policy Study (Ashley/Myer/Smith, Inc., 1973); "Regulation of Adult Businesses Through Zoning After Renton" (Stein, 1987).

13. The Supreme Court has upheld the constitutionality of anti-Skid-Row ordinances; in Detroit, for example, adult theaters are required to be more than 1,000 feet from adult bookstores, cabarets, and bars. The Detroit ordinance was passed to prevent a concentration of adult theaters that might lower property values, attract transients, and encourage the flight of other businesses. The court upheld the use of zoning in this case "as a tool to promote the legitimate interest in planning and regulating the use of property for commercial purposes" (Young v. American Mini Theatres, Inc., 1976).
14. Direct regulation of communication activities occurs when zoning laws place limitations upon the types of signs and billboards permitted in an area, thus affecting commercial expression in an effort to prevent traffic hazards and promote aesthetic concerns (Metromedia v. San Diego, 1981). In obscenity and billboard cases governmental authorities attempt to regulate in the name of public welfare but are aware of the impact their actions may have upon communicative activity.
15. There are presently ten different types of residence districts designed to create and perpetuate a specific character for each area; e.g., Rl and R2 districts are single-family detached residence districts (N.Y.C. Department of City Planning, 1988, article 2, chapter 1.).
16. Sidewalk cafes are prohibited on thoroughfares such as Manhattan's 86th Street from the East River to fifth Avenue, 34th Street and 23rd Street, East River to Eighth Avenue, 14th Street from second Avenue to eighth Avenue, and the entire length of Lexington Avenue. No enclosed sidewalk cafes are permitted in areas such as Central Park South from fifth to eighth Avenue, and Park Avenue South from 31 Street to 38th Street (N.Y.C. Department of City Planning, 1988, appendix B).
17. The exception to this is the structural members of the enclosing walls and windows.
18. Case law in New York for instance does not support exclusionary zoning.
19. New York courts have recently endorsed the use of incentive zoning to promote certain amenities as a trade off (rather than an exaction) from developers. Developers providing amenities such as low-income housing are being encouraged via incentive zoning (Asian Americans for Equality v. Koch, 1988).
20. The results of such incentives include the Seagram Building plaza and the plaza in front of 1 Chase Manhattan Plaza, the Citicorp Center, the IBM and AT&T headquarters, all in Manhattan, International Place in Boston and Fountain Square in Cincinnati (Whyte, 1988). In 1982 the New York City planning commission endorsed a sweeping revision of midtown zoning. Incentive zoning bonuses were dropped except for the creation of plazas and urban parks. The bonus for a plaza was reduced. The planning commission got tougher about amenities. Instead of giving bonuses for them, it mandated them (Whyte, 1988).

References

Alexander, C., Iskikawa, S., & Silverstein, M. (1977). *A pattern language: Towns, buildings, construction*. New York, NY: Oxford University Press.

Altheide, D. L., & Snow, R. P. (1979). *Media logic*. Beverly Hills, CA: Sage.

Anderson, R. M. (1986). *American law of zoning* (3rd ed.). Rochester, NY: Lawyers Co-operative Publishing.

Appleyard, D. (1981). *Livable streets.* Berkeley, CA: University of California Press.

Aristotle. (1988). *The politics* (T. A. Sinclair, Trans. and T. J. Saunders, Ed.). London: Penguin Classics.

Ashley/Myer/Smith, Inc. (1973). *City signs and lights: A policy study prepared for the Boston Redevelopment Authority and the U.S. Department of Housing and Urban Development.* Cambridge, MA: MIT Press.

Asian Americans for Equality v. Koch, 72 N.Y.2d. 121 (1988).

Barker, R. G. (1968). *Ecological psychology: Concepts and methods for stud ying the environment of human behavior.* Stanford, CA: Stanford University Press.

Bassett, E. (1936). *Zoning: The laws, administration, and court decisions during the first twenty years.* New York, NY: Russell Sage Foundation.

Bescher, J., Wright, R., & Gitelman, M. (1976). *Cases and materials on land use* (2nd ed.) St. Paul, MN: West Publishing.

Brill, M. (1989). Transformation, nostalgia, and illusion in public life and public place. In I. Altman & E. Zube (Eds.), *Public places and spaces: Human behavior and environment* (Vol. 10, pp. 7–28). New York, NY: Plenum.

The broadcasting yearbook (1990). Washington, DC.

Burgoon, J., Buller, D., & Woodall, W. G. (1989). *Nonverbal communication: The unspoken dialogue.* New York, NY: Harper & Row.

Carey, J. W. (Ed.). (1988). *Media, myths, and narratives: Television and the press.* Beverly Hills, CA: Sage.

Clay, G. (1987). The street as teacher. In A. V. Moudon (Ed.), *Public streets for public use* (pp. 95–109). New York, NY: Van Nostrand Reinhold.

Czarnowski, T. V. (1978). The street as a communications artifact. In S. Anderson (Ed.), *Onstreets* (pp. 207–212). Cambridge, MA: MIT Press.

Dovey, K. (1985). An ecology of place and place making: Structures, processes, knots of meaning. In K. Dovey, P. Downton, & G. Missingham (Eds.), *Place and placemaking: Proceedings of the PAPER 85 Conference* (pp. 93–110). Melbourne, Australia.

Eichner, R. B., & Tobey, H. (1987). Beyond zoning. In A. V. Moudon (Ed.), *Public streets for public use* (pp. 276–283). New York, NY: Van Nostrand Reinhold.

Euclid v. Ambler Realty Co., 272 U.S. 365 (1926).

Ex Parte Quong Wo, 161 Cal. 220 (Cal. 1911).

Federal Communications Commission. (1990). *FCC monitoring report.* Washington, DC: US Government Printing Office.

Francis, C., & Cooper, C. (1991). *Places people take their problems.* Paper presented at the meeting of the Environmental Design and Research Association, Oaxtepec, Mexico.

Francis, M. (1987). The making of democratic streets. In A. V. Moudon (Ed.), *Public streets for public use* (pp. 23–39). New York, NY: Van Nostrand Reinhold.

Gehl, J. (1987). *Life between buildings: Using public space.* New York, NY: Van Nostrand Reinhold.

Gillmor, D. M., Barron, J. A., Simon, T. F., & Terry, H. A. (1990). *Mass communication law: Cases and comment* (5th ed). St. Paul, MN: West.

Goffman, E. (1963). *Behavior in public places: Notes on the social organization of gatherings.* New York, NY: Free Press of Glencoe.
Gumpert, G. (1987). *Talking tombstones and other tales of the media age.* New York, NY: Oxford University Press.
Hadacheck v. Sebastian (1915), 239 U.S. 394.
Hall, E. T. (1966). *The hidden dimension.* Garden City, NY: Doubleday.
Hawke, D. (1966). *The colonial experience.* Indianapolis, IN: Bobbs-Merrill.
Heerwagen, J. H. (1990). The psychological aspects of windows and window design. In R. Selby, K. Anthony, J. Choi, & B. Orland (Eds.), *Coming of age: Proceedings of the Environmental Design Research Association annual conference* (Vol. 21, pp. 269–280).
Hiss, T. (1990). *The experience of place: A completely new way of looking at and dealing with our radically changing cities and countryside.* New York, NY: Alfred A. Knopf.
Jackson, J. B. (1987a). Forum follows function. In N. Glazer & M. Lilla (Eds.), *The public face of architecture: Civic culture and public spaces* (pp. 117–123). New York, NY: Free Press.
Jackson, J. B. (1987b). The discovery of the street. In N. Glazer & M. Lilla (Eds.), *The public face of architecture: Civic culture and public spaces* (pp. 75–86). New York, NY: Free Press.
Jacobs, J. (1961). *The death and life of great American cities.* New York, NY: Random House.
Kelly, E. D., & Raso, G. J. (1989). Sign regulation for small and midsize communities (No. 419). Planning Advisory Service Report.
Klapper, J. T. (1960). *The effects of mass communication.* New York, NY: Free Press.
Latané, B. (1981). The psychology of social impact. *American Psychologist, 36*, 343–356.
Lockridge, K. A. (1970). *A New England town: The first hundred years.* New York, NY: Norton.
Mandelker, D., & Cunningham, R. (1985). *Planning and control of land development: Cases and materials* (2nd ed). Charlonesville, VA: Michie.
Mandelker, D., & Ewald, W. R. (1988). *Street graphics and the law.* Washington, DC: American Planning Association.
Manheim, J. B. (1987). A model of agenda dynamics. In M. L. McLaughlin (Ed.), *Communication Yearbook 10* (pp. 499–516). Beverly Hills, CA: Sage.
McCombs, M. E., & Shaw, D. L. (1972). The agenda-setting function of mass media. *Public Opinion Quarterly, 36*, 176–187.
McKinney's consolidated laws of New York annotated: With annotations from state and federal courts and state agencies. (1987). *Penal Code §240.36 & 240. 21.* St. Paul, MN: West.
McQuail, D. (1983). *Mass communication theory: A introduction.* Beverly Hills, CA: Sage.
Mehrabian, A. (1976). *Public places and private spaces: The psychology of work, play and living environments.* New York, NY: Basic Books.
Metromedia, v. San Diego, 453 U.S. 490 (1981).
Metzenbaum, J. (1955). *The law of zoning* (2nd ed). New York, NY: Baker, Voorhis.
Meyrowitz, J. (1985). *No sense of place: The impact of electronic media on social behavior.* New York, NY: Oxford University Press.
Middleton, K. R., & Chamberlin, B. F. (1988). *The law of public communication.* White Plains, NY: Longman.
Moore, A., & Sullivan, J. P. (1988, June 26). Putting the old neighborhood in the new. *New York Times*, p. 30.

Mumford, L. (1987). The medieval town. In N. Glazer & M. Lilla (Eds.), *The public face of architecture* (pp. 60–74). New York, NY: Free Press.
New York City Department of City Planning. (1988). *Zoning resolution*. New York, NY: Author.
Oldenburg, R. (1989). *The great good place: Cafes, coffee shops, community centers, beauty parlors, general stores, bars, hangouts and how they get you through the day*. New York, NY: Paragon House.
Pember, D. (1990). *Mass media law* (5th ed). Dubuque, IA: Wm. C. Brown.
Perin, C. (1977). *Everything in its place*. Princeton, NJ: Princeton University Press.
Pool, I.de S. (Ed.). (1977). *The social impact of the telephone*. Cambridge, MA: MIT Press.
Rudofsky, B. (1969). *Streets for people; a primer for Americans*. Garden City, NY: Doubleday.
Rykwert, J. (1978). *The street: The use of its history*. In S. Anderson (Ed.), *On streets* (pp. 15–28). Cambridge, MA: MIT Press.
Schneekloth, L. H. (1987). Current forms of practice and application. In E. Zube & G. T. Moore (Eds.), *Advances in environment, behavior and design* (Vol. 1, pp. 307–334). New York, NY: Plenum.
Schramm, W. (1973). *Men, messages, and media: A look at human communication*. New York, NY: Harper & Row.
Sennett, R. (1974). *The fall of public man: On the social psychology of capitalism*. New York, NY: Vintage Books.
Sennett, R. (1987). The public domain. In N. Glazer & M. Lilla (Eds.), *The public face of architecture: Civic culture and public spaces* (pp. 26–47). New York, NY: Free Press.
Sennett, R. (1990). *The conscience of the eye: The design and social life of cities*. New York, NY: Knopf.
Sharp, W. (1989). Special permits and site plan approval. In *Basics of zoning and land use control* (pp. 53–68). Albany, NY: New York State Bar Association.
Shaw, D. L., & McCombs, M. E. (1977). *The emergence of American political issues: The agenda-setting function of the press*. St. Paul, MN: West.
Stein, R. M. (1987). Regulation of adult businesses through zoning after Renton. *Pacific Law Journal, 18*, 351.
Sweet, J. (1987). *Legal aspects of architecture, engineering and the construction process* (2nd ed.). St. Paul, MN: West.
Town of Oyster Bay Building Zone Ordinance. (1989).
Vidler, A. (1978). The scenes of the street: Transformations in ideal and reality, 1750–1871. In S. Anderson (Ed.), *On streets* (pp. 29–111). Cambridge, MA: MIT Press.
Village of Great Neck, §210–11; §210–50 and 210–63.
Village of Great Neck Plaza. (1989). Local zoning law amendments.
Webber, M. M. (1964). The urban place and the nonplace urban realm. In *Explorations into urban structure* (pp. 79–153). Philadelphia, PA: University of Pennsylvania Press.
Welch v. Swasey (1909).
Whyte, W. H. (1988). *City: Rediscovering the center*. New York, NY: Doubleday.
Willis, H. (1989). Zoning board of appeals appellate jurisdiction. In *Basics of zoning and land use control* (pp. 45–50). Albany, NY: New York State Bar Association.

Yokley, E. C. (1978). *Zoning law and practice* (4th ed.). Charlottesville, VA: Michie.
Young v. American Mini Theatres, Inc., 427 U.S. 50 (1976).

References, Introduction 2016

Dunlap, D. W. (2016, July 25). Zoning arrived 100 years ago. It changed New York City Forever. *New York Times.* Retrieved from http://www.nytimes.com/2016/07/26/nyregion/new-yorks-first-zoning-resolution-which-brought-order-to-a-chaotic-building-boom-turns-100.html?_r=0

Eide, S. (2016, July 8). Next City. Retrieved from https://nextcity.org/daily/entry/new-york-1916-zoning-law-setback-architecture-design

Fox, J. (2016, July 27). Zoning has had a good 100 years. Enough already. *Bloomberg View.* https://www.bloomberg.com/view/articles/2016-07-27/zoning-has-had-a-good-100-years-and-that-s-plenty

Graham, V. (2016, July 25). 100 years after zoning in New York City, government dominates land use. *Market Urbanism.* Retrieved from http://marketurbanism.com/2016/07/25/100-years-after-zoning-in-new-york-city-government-dominates-land-use/

Gumpert, G., & Drucker, S. J. (2010). Media and the archaeology of the city: Historical perspectives on urban and media developments. In H. Jassem, S. J.Drucker, & G. Burd (Eds.), *The urban communication reaader: Volume Two.* Cresskill, NJ: Hampton Press.

Weisbrod, C. (2016). ZONING @ 100 Meditations on the 100th Anniversary of New York City's zoning resolution. Retrieved from http://www.zoning100.com/#!weisbrod-100-years-of-zoning/z1csc

Zoning NYC Scavenger Hunt: Celebrating 100 Years of NYC Zoning. (2016, August 2). GKV Architects. Retrieved from http://www.gkvarchitects.com/journal/zoning-nyc-scavenger-hunt-celebrating-100-years-of-nyc-zoning/

CHAPTER SIX

Limiting Participatory Culture

The New Police Power and the Legitimization of Free Speech Zones

DAVID S. ALLEN
University of Wisconsin-Milwaukee

Introduction

The regulatory structure that shapes the American idea of expressive freedom is often at odds with its philosophical roots. Nowhere is this more evident than in the management of public expressive activities.[1] If it is assumed that public space (either real or virtual) is essential for citizens to participate in democracy, then the regulation of that space is of vital interest for any society that aspires to be democratic (Kalven, 1965).

In recent years, numerous examples in the United States have illustrated the regulation-philosophy problem, but perhaps none illustrate that contradiction as concisely as does the concept of free speech zones, where protesters are locked behind barricades or chain-link fences. How can a society that prides itself on having the greatest expressive freedom in the world, and that presumably recognizes the importance of that freedom to democracy, justify confining those who engage in political dissent?

This chapter attempts to better understand that apparent contradiction by examining the history and development of the regulatory structure that has legitimized free speech zones. It will be argued that the key to understanding much of the regulation of public expressive acts in the United States can be traced to the changing fortunes of the often-ignored police power—the idea that government is allowed to make regulatory decisions that promote the public welfare. The police power not only established a legal regime that legitimized the zoning of public space, but it also cleared the way to allow governmental officials to make moral choices about what types of actions served the public interest. While scholars have recognized the police power's importance for understanding the regulation of public expressive acts in pre-1920s America (see Pfohl, 1993; Rabban, 1997), today it is generally dismissed as having little relevance to First Amendment issues (Schiller, 2000, p. 75). However, the regulatory structure that legitimizes today's free speech zones has important ties to the police power.

After reviewing the rise and fall of the police power as a way of regulating expression in 20th-century America, it will be suggested that much of today's regulatory structure of public expressive acts is built on what I term the new police power. This new police power, built on protecting individual freedom behind a façade of objective legal rules and tests, masks a philosophy of public life that separates the regulation of expressive freedom from its philosophical roots. Rather than viewing public space as a place for citizens to come together and share ideas, the new police power serves to discourage participation and puts forward a picture of dissent as something to be feared and controlled. As such, the new police power not only legitimizes free speech zones, but it also hinders the development of a participatory, democratic culture.

Participatory Culture and Democratic Spaces

Ideas about the connection between freedom of expression, public space, and participation run deep in democratic theory. Richard Sennett, in his study of the city in western civilization, notes the importance of public squares or open spaces in ancient Greece. These public spaces allowed—in fact, encouraged—citizens to engage in what Sennett terms a form of participatory democracy, moving among different groups and exposing themselves to different ideas (Sennett, 1994, p. 55).

John Stuart Mill noted the importance of public expressive acts and their importance to democracy, not only in his philosophical writings but also in his work as a minister of parliament. In 1867, Mill addressed Parliament in opposition to a bill that would prevent public meetings in royal parks. A citizen, Mill said, "has a right to be, there, according to the Constitution of this country, he has a right to talk politics, to one, to fifty, or to 50,000 persons. I stand up for the right of doing this in the parks" (quoted in O'Rourke, 2001, p. 152). The purpose of these public expressive acts was not simply as an aid to discussion, but as a form of "demonstration" (O'Rourke, 2001, p. 152).

While Mill was concerned with real, tangible public space, theorists have long recognized the importance of protecting the space that emerges from the interactions of citizens, what today is often termed virtual space. Hannah Arendt's notion of the *polis* sought to protect the space that "arises out of acting and speaking together," no matter where people might be physically located (Arendt, 1958, p. 198). In that same vein, Jürgen Habermas's notion of the public sphere seeks to protect spaces that emerge from the communicative actions of citizens (Habermas, 1996, p. 360). These ideas are at the heart of today's theorizing about the space that citizens create through their use of technology (see Kitchin & Dodge, 2011).

In the United States, much of the discussion about public expressive acts has been tied to the idea of the public forum. That connection was first made by the U.S. Supreme Court in the 1939 case *Hague v. CIO*, when Justice Owen Roberts wrote:

> Wherever the title of streets and parks may rest, they have immemorially been held in trust for the use of the public and, time out of mind, have been used for the purposes of assembly, communicating thoughts between citizens, and discussing public questions. (*Hague v. CIO*, 1939, p. 515)

Connections between the public forum and democratic theory were expanded in an influential 1965 law review article by Harry Kalven, Jr., shaping the contours of the modern American idea of the public forum and tying the public forum directly to democracy.

> [I]n an open democratic society the streets, the parks, and other public places are an important facility for public discussion and political process. They are in brief a public forum that the citizen can commandeer; the generosity and empathy with which such facilities are made available is an index of freedom. (Kalven, 1965, pp. 11–12)

Despite that literature, the protection of public space for democratic purposes has enjoyed a checkered history. Fears of unions, communists, and terrorists have led to numerous attempts by government to curtail a variety of public activities (see Zick, 2009). Zoning laws and property rights, including copyright (see Boyle, 2008) and its regulation of the use of digital public space (see Balkin, 2004), have been enlisted in an attempt to make public life more orderly and efficient. And in that search for order, theorists have debated what kind of participation is required for a democracy to thrive (see Held, 1987).

The connection between participation, freedom of expression, and public space in the digital age has been attracting increased attention. Jack Balkin has argued that the purpose of freedom of speech is to promote democratic culture—the "individual's ability to participate in the production and distribution of culture" (2004, pp. 3–4). As a result, expression is "both individual and cultural" (2004, p. 5) and protects the individual's ability to participate in democracy. Balkin notes that newer, digital technologies can both limit and expand participatory potential. He stresses the need to understand the infrastructure that supports freedom of expression, making sure that it protects the participatory role of democracy (Balkin, 2004, p. 6).

While Balkin was mainly concerned with the rise of digital technologies, his writings point to the need to understand how the infrastructure that governs public space can work to impede participation within society. Understanding that spatial infrastructure also helps us understand the rise of free speech zones within society. The use of free speech zones can be traced to the late 1960s as U.S. colleges and universities sought a way to control dissent during the Vietnam War. The first recorded legal challenge involving a free speech zone arose out of a protest at Southwest Texas State University. In *Bayless v. Martine*, students challenged the creation of that "student expression area" as a restriction on their First Amendment rights. A court eventually upheld the constitutionality of the zone, noting that the expression area allowed students to use the zone during certain periods of time but did not exclude protests in other areas of campus (see *Bayless v. Martine*, 1970).

Free speech zones became more popular in the 1980s, with the National Park Service developing rules for the establishment of areas available for "public assemblies" within national parks (Demonstrations, 2012). In 1988 a free speech zone was used for the first time at a national political convention, with one being established in Atlanta, Georgia, for the Democratic National Convention. While no legal challenges emerged from the Atlanta convention, challenges did arise with the use of free speech zones at Democratic Party conventions in 2000 in Los Angeles, 2004 in Boston, and 2008 in Denver, Colorado. In general, courts rarely questioned the government's ability to control public space, instead turning their attention to the siting and aesthetics of the zones themselves.

Today the zoning of public spaces has become accepted practice. While government seeks to control protesters through the use of zones, protesters themselves establish zones to carve out protected space, free from the influence of government and media. Free speech zones, occupy movements, and safe-space zones dominate public life. And while these zones have become commonplace, their historical development and their impact on democracy remains relatively unexamined.

The Police Power and Expressive Freedom

Prior to the invention of the modern understanding of the First Amendment in the 1930s, with its emphasis on negative liberties and civil liberties, the police power dominated thinking about expressive freedom (Schiller, 2000, p. 11). As Richard T. Pfohl has suggested, early discussions about expressive freedom and public space did not center on the First Amendment, but rather on the limits of governmental power (Pfohl, 1993, p. 538).

Under the police power, when faced with questions about expressive freedom, officials were to make decisions based on what would protect the "public safety, order, or morals" of society (Schiller, 2000, p. 12). As a result, expressive freedom was not simply a legal question, but a moral question that allowed government to make assumptions about what actions might benefit the public welfare.

The Rise of the Police Power and the Spatiality of Expression

Much of our understanding of the police power can be traced to Ernst Freund's seminal 1904 book, *The Police Power*. In the book Freund not only sought to articulate a new role for the police power to play in democratic society, but to also capture its development through an examination of the case law. The book gives birth to the administrative law movement in the United States (Kraines, 1974, pp. 10–11) and promotes the development of zoning laws as a solution to public problems (see Toll, 1969). As such, Freund's work is an early exploration of the spatiality of expressive freedom.

Freund was, of course, not the first to write about the police power. Santiago Legarre suggests that while the term was popular in American constitutional law since being coined by Chief Justice John Marshall (see *Brown v. Maryland*, 1827), "it is one of the most misunderstood ideas in constitutional law" (Legarre, 2007, p. 747). While the term "police power" is not used in the U.S. Constitution, over the years the police power has been carved out by relying on Article I, Section 8[2] of the Constitution and the Tenth Amendment[3] to the Constitution.

Thomas M. Cooley appears to be one of the first to connect the police power and expressive freedom. In his 1868 work, *A Treatise on the Constitutional Limitations*, Cooley, in concluding a chapter on the state police powers, writes that "the state has also the authority to make extensive and varied regulation as to the time, mode, and circumstances in and under which parties shall assert, enjoy or exercise their rights …" (Cooley, 1972, p. 597). In his discussion of the police power and its influence on expressive acts, Cooley devoted most of his time to demonstrating how it can be used to limit obscenity. Cooley said little about the state's ability to regulate expressive acts in public locations such as streets and parks. He did note that streets are constructed by the state and therefore it has "full power to provide all proper regulation of police to govern the action of persons using them" (Cooley, 1972, p. 588), but he never addressed the rights of citizens to use streets for expressive purposes.

As a result, it was left to Freund to make the connection between the police power and expressive activity. Daniel Ernst has argued that Freund's goal in much of his legal work was to transfer to the United States the German idea of *Rechtsstaat*, the idea that the polity was able to clearly define and limit the activities of the state. Freund admired Europe's ideas about city planning, social insurance, and rural cooperatives (Ernst, 2009, p. 178). But while he saw many advantages to those ideas, he also worried about a strong administrative state interfering with individual liberties.

Freund divided the police power into three spheres of "activities, conditions and interests" (Freund, 1976, p. 11). First was the conceded sphere. Covered by what Freund termed an "ever increasing amount of restrictive legislation," it covered a wide range of safety, order and morals within society. Second was the debatable sphere applying to "proper production and distribution of wealth." Freund viewed legislation in this sphere as being in the experimental stage. And third, the exempt sphere of "moral, intellectual and political movements" that was related to individual liberty, including expressive freedom (Freund, 1976, p. 11). He was quick to note, however, that the spheres often overlap forcing the state to reconcile competing interests (Freund, 1976, p. 11).

Freund recognized expressive liberty as an individual freedom, but questioned how far that freedom extended in public settings. Freund finds protection for "social intercourse" that would not "endanger public safety, health, order or morality" (Freund, 1976, p. 487). His ideas are most clearly laid out in his thinking about the right of assembly. Recognizing that the right to meet for the purpose of "debate, discussion or cooperation" is constitutionally protected, Freund also notes that society has a right to interfere when actions cross into unlawfulness (Freund, 1976, pp. 515–516). That reasonable regulation, in Freund's words, involved regulations as to "time and place, and number and duration" of expressive activities (Freund, 1976, p. 515).

Streets and other public places owned and controlled by government are subject to the police power, he noted, but he differentiated between what constituted a reasonable police-power action and ownership. Freund was critical of the idea, as put forward by the U.S. Supreme Court in the late 1800s, that parks are owned by government in the same way that private people own their houses (see *Davis v. Massachusetts*, 1897).

For Freund, expressive activity needed to be protected from "impairment or suppression," but could be subjected to "reasonable regulation" that would protect values identified in the police power (Freund, 1976, p. 521).

For Freund, individual expression was protected up to the point where it might have some negative impact on society. At that point, the police power becomes a viable option for controlling public expressive acts that might tend to have a negative impact on society. In the words of the dominant free speech test of the period, expression that might have a bad tendency for society could be prohibited (see Rabban, 1997).

The New Deal and the Decline of the Police Power

While Freund's ideas about administrative law and the police power were the foundation of the regulation of expressive freedom prior to the 1930s, New Deal-era reformers broke from Freund in an attempt to articulate a clearer understanding of government's role in the regulation of public space. Whereas Freund sought to use administrative law and the police power as a way of limiting government, New Deal reformers sought to use the police power as a way of empowering government to make decisions in the public's interest.

The debate over the role of the police power in society was part of a wide-ranging societal debate among different groups in the early- to mid-1900s, often dividing progressives and pragmatists. Broadly speaking, progressives, such as Freund, were more likely to see a positive role for government and elites to play in the management of public life, while pragmatists tended to be more suspicious of that governmental role. For the progressives, the goal was designing administrative agencies that would use the police power wisely and ethically to improve society. For the pragmatists, the police power reserved far too much power to the government and limited the formation of a true public sphere.

The debate between progressives and pragmatists was played out in a series of expression cases that reached the U.S. Supreme Court, stretching from *Near v. Minnesota* to *Hague v. CIO.* And as the *Near* decision came before the U.S. Supreme Court in 1931, the pragmatists were winning. As Reuel Schiller has argued, the *Near* decision came at a time of growing skepticism about not only the police power, but also about the ability and wisdom of putting too much power in the hands of administrative experts. The cure in the eyes of many was giving the courts more power over public life (Schiller, 2007, p. 423).

By the end of the 1930s, the break from the police power in the area of the expressive freedom would be complete and in *Hague v. CIO* the court, through Justice Roberts, planted the seeds of the idea of the public forum. The *Hague* decision built on a view of society heavily influenced by pragmatism—an influence that becomes clear by examining the work of the American Bar Association's Committee on the Bill of Rights. The committee's influential friend-of-the-court brief in *Hague* makes it clear that the purpose of the public forum is its social nature—its ability to bring a diverse citizenry together into a cohesive public (see Allen, 2014, pp. 145–146).

The break from Freund's idea of the police power was made clear later in 1939 when the Court ruled that citizens have a right to use city streets for expressive purposes. In *Schneider v. State* the Court refused to allow cities to use the police power to stop citizens from distributing circulars on streets and in other public places. As Justice Roberts wrote in his majority opinion:

> [T]he streets are natural and proper places for the dissemination of information and opinion; and one is not to have the exercise of his liberty of expression in appropriate places abridged on the plea that it may be exercised in some other place. (*Schneider v. State*, 1939, p. 163)

The Court's desire to limit governmental power did not sit well with New Deal-era progressives. Much of this dispute was played out in an intellectual debate about the meaning and purpose of administrative law. It pitted Freund against Felix Frankfurter, then a Harvard University law professor and soon to become a U.S. Supreme Court justice. As captured by Daniel R. Ernst, both Freund and Frankfurter were commissioned by the Legal Research Committee of The Commonwealth Fund (Ernst, 2009, pp. 173–176) in the early 1920s to study the expansion of administrative agencies in the United States. Joined by Harvard Law School Dean Roscoe Pound, a member of the Legal Research Committee, the work would highlight the important distinctions within progressivism itself and in its move away from pragmatism.

Ernst notes that after the battle, Freund's ideas about finding ways to limit administrative discretion were largely forgotten in favor of Frankfurter's ideas (2009, pp. 185–186). Administrative law was no longer viewed as a way to control government, but as an avenue to allow government to make the moral choices essential for the proper functioning of democracy. In the area of expressive freedom, the problem facing New Deal progressives was that prior judicial decisions, stretching from *Near* through *Hague*, had removed the police power as a way of managing public life. The question increasingly became how courts might make police-power decisions without using the police power?

The New Police Power: The Foundation of Free Speech Zones

In the 1960s and 1970s a new theory of public space began to emerge. With the U.S. Supreme Court rejecting the police power in the 1930s as an acceptable way to limit expression, a new way of regulating expression was needed. So complete was this rejection that courts today rarely refer to the police power in expressive freedom cases. Still, the new police power differs from the older police power in both its source and form.

By the 1960s, favoring a view of expressive freedom based on individual rights as opposed to societal needs, courts began fashioning new standards that would guide governmental regulation of public expressive acts. These new standards, however, did not remove government from making value choices. Rather, they merely moved institutional authority to a different branch of government. The standards established under the new police power increasingly became judicial standards that came to be adopted by local, state, and federal government.

While older notions of the police power were explicit about the moral choices that government made and the role that government played in protecting the public welfare, the decisions made under the new police power were far less transparent. Today's value choices are hidden behind a layer of "objective" standards and tests that purport to treat individuals equally. The three tests that are most important to the establishment of free speech zones are the public forum doctrine; the time, place, and manner test; and the ample alternatives test. These tests, central to the new police power, cleared the way for government to claim that decisions about public space were policy decisions rather than moral choices. But just as theorists argue that space is never neutral (Harvey, 1990, p. 239), so are the rules that govern the use of that space, hiding moral choices within a highly rationalized system that often defy both understanding and comprehension.

The Public Forum Doctrine and the Categorization of Public Life

The vision of public space that began to emerge in the 1960s bore little resemblance to the public forum of the 1930s. While early ideas about the public forum were built on discussion and a place for citizens to come together, by the 1960s those ideas had been subsumed by the rush for individualism. One historian has summed up the 1960s and 1970s as a time when society began moving away from integration and toward diversity, where the nation became less a melting pot and more a cluster of discrete peoples and cultures (Schulman, 2001, pp. 68–71).

The move toward the individualization of public life required a legal framework to support it. And in the 1960s, the U.S. Supreme Court began that move with the categorization of public space and public life. A powerful influence was the Court's decision in *New York Times v. Sullivan*, often read as simply a case about libel and press freedom. The *Sullivan* decision's impact on public life lies less in what it said about dissent and more in the structure of public life put forward by Justice William Brennan's opinion. While his call for "uninhibited, robust and wide-open" speech on "public issues" was a powerful call to protect dissent (*New York Times v. Sullivan*, 1964, p. 270), at times that statement is difficult to square with the Court's future actions.

Much of this tension can be traced to Justice Brennan's categorization of public life. He begins in *Sullivan* by limiting the decision to "public officials" and speech about "official conduct"—terms that he refused to define in any specificity (*New York Times v. Sullivan*, 1964, p. 284)—and starts the Court on its journey toward not only the categorization of libel plaintiffs but the categorization of other areas of expressive freedom.

The role categorization plays in democratic life was not a new idea in 1964. A wide range of scholars from Immanuel Kant to Max Weber to Michel Foucault (see Jones, 2009) have noted the importance of categorization and have helped us understand the paradox it presents for democratic life. Categories help us not only understand the world, but they also serve as instruments through which power is exercised and established. For the Court at the time of the *Sullivan* decision, categorization was seen as a way of insulating free-speech decisions from the vagueness of balancing tests (Ely, 1975, p. 1500) and limiting governmental power over individual freedom.

It wasn't long before the categorization introduced in *Sullivan* began to spread to the management of other areas of public life. Specifically, the Court began to categorize the public forum. Beginning in 1972, the Court began the categorization of public space, a movement that eventually evolved into today's public forum doctrine. The U.S. Supreme Court formally began the categorization of space in 1972 in *Police Department of Chicago v. Mosely*, but the basic categories that exist today were formalized in 1983 in *Perry Educational Association v. Perry Local Educators' Association*. Those categories are: traditional public forum (a place that is by tradition or government fiat open to public assembly and debate), a limited public forum (government space open to some members of the public to discuss a limited range of topics), and a non-public forum (government-controlled property that is not open to the public) (see *Cornelius v. NAACP Legal Defense and Educational Fund*, 1985).

The public forum doctrine continued to develop with multiple permutations until expressive freedom today is largely determined by the type of space people use for expressive purposes—a doctrine that, as Robert Post has noted, is "virtually impermeable to common sense" (Post, 1987, p. 1715). The categorization of the public forum has led to public space being dominated by zoning rules and spatial regulations and filled with "free speech zones," "cages," "pens," and "buffer zones" (see Zick, 2009). These areas tend to paint a picture of democracy and dissent as being an abhorrent, fearful activity, something with which citizens are not invited to engage.

Time, Place, and Manner and Free Speech Zones

The categorization of space, however, is only part of the new police power. The control of dissent in public space also required a way to manage the time and the manner in which expressive activities are done. This came with the redefinition of the time, place, and manner test. The first use of the time, place, and manner test by the U.S. Supreme Court came in the 1940 case *Cantwell v. Connecticut*, only a year after the Court first articulated its idea of the public forum. In those early years, the time, place, and manner test was primarily viewed as articulating the role that government played in managing the use of public space. As Post terms it, the time, place, and manner test played a "coordinating function" in those early years, allowing government to regulate public space in the best interests of all who wanted to use it (Post, 1995, p. 1261). The Court only saw fit to use the time, place, and manner test when there were valid competing interests over the use of public space. The time, place, and manner test was viewed as a way of equitably resolving those competing interests.

However, by the 1970s that early history and usage of the time, place and manner test had been forgotten. Rather than protecting government's coordinating function, the Court increasingly began to "subordinate" speech to valid competing state interests (Post, 1995, p. 1261). Rather than using the test as a way of coordinating public activities among competing groups, courts began to use the test as a way of empowering government to make choices about good and bad public expressive acts.

For example, in the 1976 case *Buckley v. Valeo*, one of the first cases to come before the Court concerning political campaign financing, it ruled that time, place, and manner restrictions could be applied as long as they "do not discriminate among speakers or ideas, in order to further an important governmental interest unrelated to the restriction of communication" (*Buckley v. Valeo*, 1976, p. 18).

The Court went a step further in 1984 when, in *Clark v. Community for Creative Non-violence*, it formally redefined the meaning of time, place, and manner:

> [Restrictions] are valid provided that they are justified without reference to the content of the regulated speech, that they are narrowly tailored to serve a significant governmental interest, and that they leave open ample alternative channels for communication of the information. (*Clark v. Community for Creative Non-violence*, 1984, p. 293)

While the Court's decisions from *Near* through *Hague* sought to limit the ability of government to make moral decisions about public expressive acts through the police power, the redefinition of the time, place, and manner test created a new opening for government. No longer was government to simply play a coordinating role in public life, but government was now allowed to make decisions about the morality of public expressive acts as long as it could be proven that a significant governmental interest existed.

Free Speech Zones as an "Ample Alternative"

Part of the moral choice that is buried within the time, place, and manner test is the decision about the siting of protests. The new police power, through the time, place, and manner test, gave government freedom to relocate expressive activities as long as an "ample alternative" was made available.

Being artificial spaces that lack permanence or history (that is, spaces that have been created by government to temporarily serve a narrowly defined purpose), free speech zones rely on boundaries to separate and contain citizens and the idea of ample alternatives to justify their existence. This containment allows government officials to claim they are preserving expression by allowing citizens to express themselves while at the same time making the surveillance and management of dissent more efficient. Courts have generally categorized free speech zones as limited, not traditional, public *fora* where government is given some latitude in overseeing expressive activities (see *Boardley v. U.S.*, 2010). However, complicating the issue is that the free speech zones are often carved out of space that at different times would be considered a traditional public forum (parks, parking lots, streets, etc.). Courts tend to focus their review of free speech zones on their siting and how boundaries are established. Locating zones too far away from an intended audience reduces the power of dissent (see *Members of City Council v. Taxpayers for Vincent*, 1984), while locating zones too close to an intended audience is seen as being disruptive to either the public order or the event (see *Menotti v. City of Seattle*, 2005).

As such, courts have struggled to define what constitutes an ample alternative. When the National Park Service created free speech areas on public lands, formulating rules that allowed the park service to close the areas under certain circumstances (Demonstrations, 2012), questions arose about the location of those areas. In 2004, a federal appeals court found that the Park Service had exceeded its authority by telling protesters at San Francisco Presidio National Park that they could only protest in a designated zone located some 150 to 175 yards from their preferred spot (see *Galvin v. Hay*, 2004). The court avoided the more difficult question of whether the free speech zone itself was constitutional (*Galvin v. Hay*, 2004, p. 756).

Courts have confronted free speech zones more directly in questions concerning protests at political conventions. At the 2000 Los Angeles Democratic National Convention, government designated a "secured zone" surrounding the Staples Center, the site of the convention. The area encompassed more than 8 million square feet (about 185 acres), including streets, sidewalks and other buildings (see *Service Employee International Union v. Los Angeles*, 2000). After the government established a free speech zone about 260 yards from the Staples Center, a number of groups claimed that the secured zone, as well as the free speech zone, violated their expressive rights.

At the heart of the decision by U.S. District Judge Gary Allen Feess was whether the free speech zone allowed adequate expressive opportunities to protesters. Feess found that delegates arriving at the Staples Center would have a difficult time seeing or hearing protesters due to a large media village separating protesters from delegates (*Service Employee International Union v. Los Angeles*, 2000, p. 972). Despite the fact that there was general agreement that the protest was occurring on space that would be considered to be a traditional public forum, both sides ultimately accepted a secured zone with concrete barriers and chain-link fencing closer to the delegates as an ample alternative (Barrett, 2000).

Four years later, government officials had learned from the mistakes of Los Angeles. At the Democratic National Convention in Boston, Massachusetts, government officials created two different zones: a hard security zone and a soft security zone. The hard security zone, which encompassed the FleetCenter where the convention would be staged, was under the control of the U.S. Secret Service. Only people with credentials were allowed inside the hard security zone. The soft security zone, in general surrounding and adjacent to the hard security zone, was under the control of city officials and open to all citizens (*Coalition to Protest the Democratic National Convention v. Boston*, 2004, p. 66).

The free speech zone was located within the soft zone. The zone was described by Judge Douglas P. Woodlock as a "grim, mean and oppressive space" (*Coalition to Protest the Democratic National Convention v. Boston*, 2004, p. 67). He noted that while the space was far from being ideal (*Coalition to Protest the Democratic National Convention v. Boston*, 2004, p. 77), it did provide "the only available location providing a direct interface between demonstrators and the area where delegates will enter and leave the FleetCenter" (*Coalition to Protest the Democratic National Convention*, 2004, p. 75). On appeal, a three-judge panel noted that the free speech zone provided "an opportunity for expression within sight and sound of the delegates, albeit an imperfect one" (see *Bl(A)ck Tea Society v. City of Boston*, 2008).

The cases arising out of free speech zones in Los Angeles and Boston established that the zones, no matter their location, were spaces under the control of government. The categorization of free speech zones established that these were not traditional public forums with a continuing history and culture of free speech, but rather temporary public space that fulfilled a governmental need. As a result, courts rarely questioned the legitimacy of free speech zones as government-owned space, with their barriers and wire. Judges increasingly wrestled with the location of the zones and the sanitary conditions inside the zones as they struggled to determine what constituted an ample alternative (see *American Civil Liberties Union of Colorado v. Denver*, 2008, p. 1154; *Coalition to Protest the Democratic National Convention v. Boston*, 2004, p. 68).

The new police power, as constituted through the public forum doctrine, the time, place, and manner test, and the ample alternatives test, are the foundation for the legitimation of free speech zones. This is achieved through the categorization

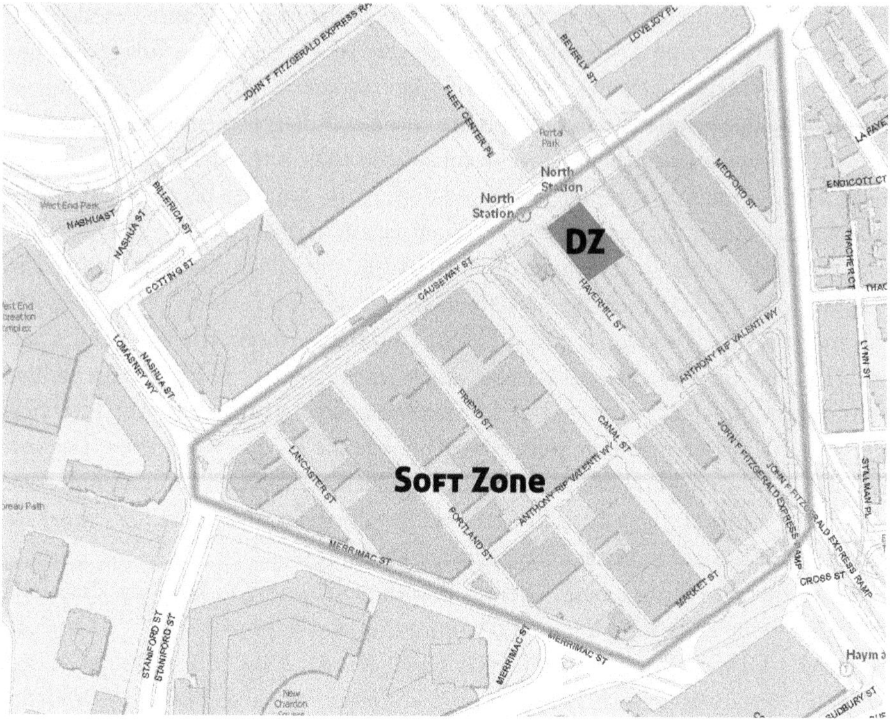

Figure 6.1. Soft zone.

of space and time, through the subordination of expressive activities to governmental interests, and to the acceptance of zones as being acceptable alternative locations for expressive activities. These decisions hide the moral choices behind a layer of policy decisions. And perhaps the most important moral choice is how free speech zones devalue democratic life and lead to even greater confusion about the role of public space and participatory culture.

The New Police Power and Participatory Culture

The new police power, with its categories, boundaries, and tests, not only has led to a confusion about the structure of public life, but has damaged participatory culture within American society. Rather than envisioning public space as a place that brings people together, the new police power has turned public space into a battleground, where citizens fight to occupy and carve out their own protected spaces.

Protests at the University of Missouri in 2015 demonstrate the corrosive impact of the new police power. During campus protests challenging the university's treatment of people of color, protesters occupied a central square. Their space was highlighted by a simple sign stating, "NO MEDIA. SAFE SPACE." As journalists attempted to enter the space, students and faculty who were occupying the space defended their right to make decisions about who was allowed to enter (Vaughn, 2015). Free-speech advocates repeatedly claimed that the protesters failed to understand the meaning of the First Amendment and public space. Once the protesters set up their camp on what was considered to be a traditional public forum, free-speech advocates argued, neither journalists nor anyone else could be excluded from entering the camp (Friedersdorf, 2015). The issue quickly changed from a protest about the rights of minority groups to the control of public space.

Of course, the new police power is not solely responsible for the confusion about public space that exists within society; other forces work to fragment and divide citizens (see Ewen, 1996). Yet it is the new police power that justifies the regulatory structure of public space today. The new police power's normative assumptions envision space as a tool to be used for controlling and managing citizens rather than an opportunity to encourage citizens to participate in public life.

That distinction is vital to sparking an active public life. Participatory culture requires more than just providing space for individuals to express their differences, it also requires common areas and a supporting culture. Free speech zones, sanctioned by the new police power, provide a certain level of individual expressive freedom, yet separate and distance citizens. To overcome this problem and aid participatory culture, regulators and citizens need to understand that at its best the public forum is not solely about meeting the expressive needs of individuals, but it is also about fulfilling a communitarian goal.

In an effort to meet those goals, regulators might rethink zoning as a positive force to aid public life. For example, Sennett has noted that when public spaces are located on the edges of communities rather than in their center, those spaces are more likely to bring diverse groups together (Sennett, 2008, p. 229). Regulators need to find ways to protect individual expression while identifying ways that will allow people to come together.

It is wrong, however, to view this problem as solely a regulatory issue. It also presents an ethical problem that is central to citizenship. Occupying public space and dividing citizens too often misses a discursive opportunity to build stronger coalitions that might challenge existing power structures. As Jürgen Habermas has noted, an active public sphere requires a supporting culture—a culture that values discourse, freedom, and participation (Habermas, 1992, p. 453). As a result, both public and private regulators, as well as citizens, need to resist actions that categorize and divide citizens.

Notes

1. The term public expressive acts is intended to capture a broad range of speech activities. As such, a public expressive act is any speech or symbolic act that utilizes public space to invite citizens to wrestle with that act's meaning.
2. Article 1, Section 8 of the U.S. Constitution states that Congress shall have the power to "provide for the common Defence and general Welfare of the United States; …" (U.S. Const. art. I, §8).
3. The Tenth Amendment reserves to the states the "powers not delegated to the United States by the Constitution" (U.S. Const. amend X).

References

Allen, D. S. (2014). The ethical roots of the public forum: Pragmatism, expressive freedom, and Grenville Clark. *Journal of Mass Media Ethics, 29*, 138–152.

American Civil Liberties Union of Colorado v. Denver, 569 F. Supp.2d 1142 (2008).

Arendt, H. (1958). *The human condition*. Chicago, IL: The University of Chicago Press.

Balkin, J. M. (2004). Digital speech and democratic culture: A theory of freedom of expression for the information society. *New York University Law Review, 79*(1), 1–58.

Barrett, B. (2000, July 29). New protest area: At Dems' doorstep. *Daily News of Los Angeles*. Retrieved September 14, 2010 from http:www.lexisnexis.com/1nacui2api

Bayless v. Martine. (1970). 430 F.2d 873 (5th Cir.).

Bl(A)ck Tea Society v. City of Boston. (2008). 378 F.3d 8 (1st. Cir.).

Boardley v. U.S. (2010). 615 F.3d 508 (D.C. Cir.).

Boyle, J. (2008). *The public domain: Enclosing the commons of the mind*. New Haven, CT: Yale University Press.

Brown v. Maryland. (1827) 25 U.S. 419.

Buckley v. Valeo. (1976). 424 U.S. 1.

Cantwell v. Connecticut. (1940). 310 U.S. 296.

Clark v. Community for Creative Non-violence. (1984). 468 U.S. 288.

Coalition to Protest the Democratic National Convention v. Boston. (2004). 327 F. Supp. 2d 61 (D. Mass.).

Cooley, T. (1972). *A treatise on the constitutional limitations*. New York, NY: Da Capo Press. (Original work published 1868.)

Cornelius v. NAACP Legal Defense and Educational Fund. (1985). 473 U.S. 788.

Davis v. Massachusetts. (1897). 167 U.S. 43.

Demonstrations, 36 C.F.R. §2.51 (2012).

Ely, J. H. (1975). Flag desecration: A case study in the roles of categorization and balancing in first amendment analysis. *Harvard Law Review, 88*, 1482–1508.

Ernst, D. (2009, October). Ernst Freund, Felix Frankfurter and the American *Rechtstaat*: A transatlantic shipwreck, 1894–1932. *Studies in American Political Development*, 23, 171–188.

Ewen, S. (1996). *PR! A social history of spin*. New York, NY: Basic Books.
Freund, E. (1976). *The police power: Public policy and constitutional rights*. New York, NY: Arno Press. (Original work published 1904.)
Friedersdorf, C. (2015, November 10). Campus activists weaponize 'Safe Space.' *The Atlantic*. Retrieved from http://www.theatlantic.com/politics/archive/2015/11/how-campus-activists-are-weaponizing-the-safe-space/415080/
Galvin v. Hay. (2004). 374 F.3d 739 (9th Cir.).
Habermas, J. (1992). Further reflections on the public sphere. In C. Calhoun (Ed.), *Habermas and the public sphere* (pp. 421–461). Cambridge, MA: MIT Press.
Habermas, J. (1996). *Between facts and norms: Contribution to a discourse theory of law and democracy* (William Rehg, Trans.). Cambridge, MA: MIT Press.
Hague v. CIO. (1939). 307 U.S. 496.
Harvey, D. (1990). *The condition of postmodernity*. Cambridge, MA: Blackwell.
Held, D. (1987). *Models of democracy*. Stanford, CA: Stanford University Press.
Jones, R. (2009). Categories, borders and boundaries. *Progress in Human Geography, 33*(2), 174–189.
Kalven, H. Jr. (1965). The concept of the public forum: *Cox v. Louisiana. The Supreme Court Review, 1965*, 1–32.
Kitchin, R., & Dodge, M. (2011). *Code/space: Software and everyday life*. Cambridge, MA: MIT Press.
Kraines, O. (1974). *The world and ideas of Ernst Freund*. Birmingham, AL: University of Alabama Press.
Legarre, S. (2007, February). The historical background of the police power. *University of Pennsylvania Journal of Constitutional Law, 9*, 745–796.
Members of City Council v. Taxpayers for Vincent. (1984). 466 U.S. 789.
Menotti v. City of Seattle. (2005). 409 F.3d 1113.
Near v. Minnesota. (1931). 283 U.S. 697.
New York Times v. Sullivan. (1964). 376 U.S. 254.
O'Rourke, K. C. (2001). *John Stuart Mill and freedom of expression: The genesis of a theory*. London: Routledge.
Perry Educational Association v. Perry Local Educators' Association. (1983). 460 U.S. 37.
Pfohl, R. T. (1993, Spring). *Hague v. CIO* and the roots of public forum doctrine: Translating limits of powers into individual rights. *Harvard Civil Rights-Civil Liberties Law Review, 28*, 533–566.
Police Department of Chicago v. Mosely. (1972). 408 U.S. 92.
Post, R. (1987). Between governance and management: The history and theory of the public forum. *UCLA Law Review, 34*, 1713–1835.
Post, R. (1995). Recuperating first amendment doctrine. *Stanford Law Review, 47*(6), 1249–1281.
Rabban, D. M. (1997). *Free speech in its forgotten years*. New York, NY: Cambridge University Press.
Schiller, R. E. (2000). Free speech and expertise: Administrative censorship and the birth of the modern first amendment. *Virginia Law Review, 86*(1), 1–102.

Schiller, R. E. (2007). The era of deference: Courts, expertise, and the emergence of new deal administrative law. *Michigan Law Review, 106*, 399–441.

Schneider v. State. (1939). 308 U.S. 147.

Schulman, B. (2001). *The seventies: The great shift in American culture, society, and politics*. Boston, MA: Da Capo Press.

Sennett, R. (1994). *Flesh & stone: The body and the city in western civilization*. New York, NY: W. W. Norton.

Sennett, R. (2008). *The craftsman*. New Haven, CT: Yale University Press.

Service Employee International Union v. Los Angeles. (2000). 144 F. Supp.2d 966.

Toll, S. I. (1969). *Zoned American*. New York, NY: Grossman Publishers.

U.S. Const. amend. X.

U.S. Const. art. I, §8.

Vaughn, J. M. (2015, November 9). Mizzou protesters: Stay out of our 'safe space' or we'll call the cops. *The Daily Beast*. Retrieved from http://www.thedailybeast.com/articles/2015/11/09/mizzou-protesters-to-media-stay-out-of-our-safe-space-or-we-ll-call-the-cops.html

Zick, T. (2009). *Speech out of doors: Preserving first amendment liberties in public places*. New York, NY: Cambridge University Press.

CHAPTER SEVEN

Street Performers, the First Amendment, and New York City's Activity Zones

JULIET DEE
University of Delaware

Introduction

This chapter examines the legal issues surrounding New York City's new law that regulates where street performers may ply their trade in Times Square. On June 21, 2016, the New York City Council passed Local Law 2016/53: "a local law to amend the Administrative Code of the City of New York, in relation to pedestrian plazas." The official summary stated that "The bill would provide the Department of Transportation the ability to designate pedestrian plazas." In response to this new law, the Department of Transportation then created teal (turquoise-colored) "activity zones" for costumed characters and desnudas to stand in as they ask tourists to pose for photos and hope for tips. The legal questions concern time-place-manner constraints in a public forum such as Times Square, and the legality of panhandling.

Background

In July 2014 a non-profit trade association called the Times Square Alliance counted 76 costumed characters trolling Times Square and 150 to 200 characters in a crowded area through which 350,000 to 450,000 people pass each day (Fertig, 2015). In July 2015 a dozen "desnudas" (topless women) appeared each day wearing

Figure 7.1. Costumed Characters. Source: https://www.flickr.com/photos/blakespot/8259056682/in/photolist-dzPPP1. Copyright © 2012 Blake Patterson, and made available under a Creative Commons License (https://creativecommons.org/licenses/by/2.0/legalcode).

nothing but thongs and body paint; some of the desnudas "accosted passers-by, grabbing their arms and urging them to stop and pose" for a photo ("Times Square topless ladies," 2015). The desnudas also "scared off families with children" (Haller, 2016). The desnudas make $10 to $20 per photo and can bring in $300 per day. The Times Square Alliance and the Broadway League blamed these characters' aggressive tactics for a downturn in theatre attendance on Broadway.

What brought the issue of street performers to lawmakers' attention in the first place was that a few of the costumed characters were behaving badly. For example, in Manhattan's Times Square, from 2013 to 2015, a Bank of America executive got into a scuffle with a topless woman when he thought she had stolen his wallet, a man dressed as Cookie Monster knocked down a two-year-old after the toddler's mother did not give him a tip, two "Statues of Liberty" got into a brawl, a man dressed as Elmo launched into an anti-Semitic tirade (he was later jailed for trying to extort $100,000 from the Girl Scouts), and men dressed as "Woody" and "Jessie" from *Toy Story*, Spiderman, Captain America and the "Naked Black Cowboy" were arrested for offenses such as groping women, punching a police officer, or harassing tourists.

The bad behavior of these costumed characters and desnudas led New York Mayor Bill deBlasio to appoint a task force in September 2015 to make recommendations on how to deal with them. New York City Police Commissioner William Bratton had recommended closing down the pedestrian plazas as a means of controlling the costumed characters and desnudas, but the members of Mayor Bill deBlasio's Task Force on Times Square were concerned that this would be too restrictive. Instead, the Task Force suggested "doubling the police presence in Times Square to about 100 officers" (New York Task Force to Suggest New Rules, 2015).

City officials had suggested that the topless women as well as the costumed characters be moved to designated "activity zones," but Mayor deBlasio's legal advisors explained that the quandary was whether the costumed characters and desnudas' activity comprised a street performance or a type of commerce that New York City can regulate. For the record, it is legal to ask passers-by for money and it is legal for women to go topless in New York City.

Beginning in June 2016, the costumed characters and desnudas have been corralled in teal "activity zones," while passers-by may walk in "pedestrian flow zones," as is discussed below.

Legal Questions

This chapter will focus on why and how New York City is trying to regulate "communication" in Times Square, and what New Yorkers and citizens of other cities can learn from the Times Square experience. After the street performers' bad behavior resulted in calls for order and regulation, city officials had to balance the competing rights of both tourists and New Yorkers to move through Times Square without being accosted against the First Amendment rights of the street performers to express themselves and to solicit tips. In other words, this chapter examines a regulatory web associated with diverse forms of street performance in New York City.

The chapter begins with a few definitions of terms.

Definitions

A *street entertainer* or *street performer* is a performer who works in a public area without a specific paying audience; he or she hopes to earn tips for such performances. Street performers hope to receive tips based on (1) a certain skill such as juggling or (2) merely their appearance (being topless with body paint and posing for photos). Street theatre is arguably the oldest form of theatre in existence. Modern street theatre has a unique role in our lives in the sense that anyone, including people who cannot afford to go to the "legitimate" theatre, can watch street performers.

A *busker* is someone who performs music in subway stations or other public places and hopes to receive tips from listeners. Although at least one legal scholar argues that "Busking combines artistic performance (usually musical) with panhandling" (Tushnet, 2012, p. 204), buskers would strongly disagree that they are panhandling because they are *passively* hoping that listeners will leave a few dollars in an open guitar case, but they are not *actively* requesting tips in face-to-face interactions.

To *solicit* means to accept or request a fee, donation, tip or any other form of compensation.

To *solicit in an aggressive manner* means to follow a person after soliciting, if this conduct is likely to cause a person to 1) fear bodily harm to himself or another, or 2) be intimidated into giving money or another thing of value, or 3) suffer unreasonable inconvenience, annoyance or alarm.

A *traditional public forum* is a place that is open to the general public, such as a street, sidewalk or public park. The right of free expression is contingent on the definition of what constitutes a public forum with regard to a speaker on the proverbial soapbox or protesters marching in a demonstration, for example.

Even a traditional public forum is subject to *time-place-manner* restrictions; this means that government officials can prohibit union members on strike from blocking intersections so as not to interfere with the flow of traffic, for example. Courts will accept time-place-manner restrictions on speech, provided that they are content-neutral.

Content-neutral restrictions regulate speech without regard to the point of view of that speech. An example would be ordinances that prohibit the use of loudspeakers in order to reduce noise, or that require street musicians to keep their sound levels under 85 decibels. For example, Rock against Racism, a group that sponsored programs of speeches and rock music, challenged New York City's guidelines that controlled the sound-level of concerts at the Naumberg Acoustic Bandshell in Central Park. Rock against Racism argued that they had a First Amendment right to increase the volume but the U.S. Supreme Court held that the guidelines were *content-neutral* because they applied to *all* performers. Justice Kennedy explained that the "guidelines continue to permit expressive activity in the bandstand, and have no effect on the … content of that expression beyond regulating the extent of amplification" (*Ward v. Rock against Racism,* 1989).

Content-neutral speech is reviewed under intermediate scrutiny rather than strict scrutiny because it is restricted based only on the time, place or manner by which the information is communicated, but not based on the content of the information.

The government may impose *narrowly tailored* restrictions on speech; this means that legislators must write laws specific enough to fulfill only the intended goals of the legislation.

In order to regulate speech, there must be a *compelling government interest,* referring to something necessary for the good of the general public. Applying this to the costumed characters in Times Square, the "compelling government interest" would be to ensure that both New Yorkers and tourists can move freely without being waylaid by the costumed characters.

If the government is regulating speech, it must allow *alternative channels of communication,* meaning that if a municipality bans yard signs, then it *must* permit flyers, telephone calls, and newspaper ads, for example.

Value of Street Performance as Communication

Urban dwellers (and tourists) who appreciate street performers explain that they love vibrant public spaces in which there is a contagious energy; indeed, some would even recognize street performance as an "institution." Some city dwellers perceive street performance as a valuable and sometimes irritating form of *persuasive communication* in an urban context. Street performers must function as "practical rhetoricians" in the sense that they must persuade passers-by to give them (1) attention, and (2) tips. In economic terms, street performance is "free-market-based" in the sense that there is competition from other street performers, and attention from passers-by is transient.

Urban scholar Jessie McCracken argues that

> Street performance is an important part of our urban street culture. … Street performance is an art form accessible to all [with no] admission fee. It brings people together in what Erving Goffman describes as a focused interaction where strangers become neighbors. … Street performers encourage individuals to see their environment in a new way. (McCracken, 2006, p. 45)

McCracken further argues that "the public and cultural aspects of the art form of street performance have been completely overlooked because the street performers are not generating any income for the city government" (McCracken, 2006, pp. 47–48).

Time-Place-Manner Regulations: Is There a First Amendment Right to Busk?

Street performance is clearly an important part of urban street culture. Just as a speaker in Hyde Park in London, England can stand on a soap box and try to persuade passers-by of his views, those who wish to express themselves in the United States can rely on the public forum doctrine; this would include the costumed characters in Times Square as well as political speakers. The public forum doctrine is a relatively recent legal concept that the U.S. Supreme Court outlined in a 1939 case:

> Wherever the title of streets and parks may rest, they have immemorially been held in trust for the use of the public and, time out of mind, have been used for purposes of assembly, communicating thoughts between citizens, and discussing public questions. Such use of the streets and public places has, from ancient times, been a part of the privileges, immunities, rights, and liberties of citizens. (*Hague v. Committee for Industrial Organizations*, 1939, p. 515)

A public forum is a place in which all expression that the First Amendment protects is permitted. Streets, sidewalks and public parks, from "time immemorial," are considered to be *traditional or quintessential public forums*. Legal scholar Timothy Zick explains that "The public forum is a place for unfettered speech with a built-in tolerance for boisterous and even offensive expression" (Zick, 2006, p. 620). Zick suggests that "speech and spatiality are closely related" (Zick, 2006, p. 638). The U.S. Supreme Court has held that "no one may insist upon a street meeting in the middle of Times Square at the rush hour as a form of freedom of speech" (*Cox v. Louisiana*, 1965, p. 536). In other words, a commuter's need to get to work will outweigh a protester's right to tie up traffic with political expression.

The courts also recognize *dedicated or limited public forums*; these comprise public property that the government has made available for expressive activity, such as a municipal building auditorium. Once such spaces are opened to the public, public forum rules go into effect, meaning that officials must grant access to the room or space on a content-neutral basis.

Not all publicly owned spaces qualify as public forums, however. The U.S. Supreme Court has ruled that an airport terminal is not a public forum, meaning that airport officials can ban speakers who are soliciting funds, although speakers are allowed to distribute literature (*International Society for Krishna Consciousness v. Lee*, 1992). The grounds of a county jail also comprise a *non-public forum*; thus, if a few hundred protesters are demonstrating outside the county jail, police can arrest them for trespassing (*Adderley v. Florida*, 1966). And military bases are also non-public forums; Justice Potter Stewart explained that military bases exist "to train soldiers, not to provide public forums [for political candidates]" (*Greer v. Spock*, 1976).

In grappling with the problem of aggressive street performers in Times Square, the New York City Council members were aware that their new regulations would have to meet Constitutional safeguards for freedom of expression. Public forums are subject to time-place-manner restrictions, but these restrictions (1) must be content-neutral, (2) must be narrowly tailored, (3) must serve a compelling government interest, and (4) must allow alternative channels of communication (*Clarke v. Community for Creative Non-Violence*, 1982; *Ward v. Rock against Racism*, 1989, p. 791).

New York City's Twentieth Century Ordinances

Before we examine specific ordinances, it would be helpful to consider a brief history of busking in New York City. Following waves of immigration in the late 1800s, it was common to see German bands and Italian hurdy-gurdies: "As late as 1923 [there were] 800 organ grinders and an equal number of itinerant street musicians" (Campbell, 1981, p. 11). In the 1920s, blacks migrated to northern cities seeking employment, but due to racial discrimination, there was a sharp increase in the number of black musicians performing in the streets (Tanenbaum, 1995, p. 38).

In the late 1930s during the Great Depression, many people turned to busking as their only source of income. Buskers would not have considered themselves to be *panhandlers*; as is stated above, whereas street musicians or jugglers would perform and then hope that the audience would leave money in an open violin case, panhandlers would engage people in conversation and ask for money without attempting any type of performance. But territorial disputes among buskers, merchants and vendors often resulted in violent confrontations.

New York Mayor Fiorello LaGuardia defended the poor but disapproved of beggars and panhandling. Arguing that radio and regular concerts provided enough entertainment, LaGuardia placed a ban on granting licenses for street musicians, effectively making them outlaws. When reporters asked why he had banned busking, he said, "I'm not going to license begging" (Zack, 2015, p. 68). Some historians have suggested that LaGuardia banned street music to combat Italian stereotypes associated with organ grinders, but researcher Robert Hawkins argues that LaGuardia's motivations resulted from his desire to push a work ethic: "Street musicians occupied an ambiguous space between begging and self-employment and … confusion between honest unemployment and willful dependency was a political liability for the Mayor" (Hawkins, 2012).

The ban was unevenly enforced, and at first some judges who thought it was too harsh let buskers off with only a warning (Zack, 2015, p. 68). LaGuardia's ban did not extend to political speech; strangely, no one challenged the ban on busking in court. Despite LaGuardia's regulations, in the 1940s, Woody Guthrie, Pete Seeger, and others involved in the growing urban folk revival movement pulled out guitars while waiting for their trains, in effect reclaiming public space (City Lore, 2015). A few decades later, poet Allen Ginsberg challenged the ordinance against busking and prevailed on First Amendment grounds; in response, Mayor John Lindsay lifted the ban on busking in the 1970s.

Former Mayor Rudolph Giuliani "made his name cleaning up Times Square's seedy triple X-rated movie theaters. His successor, Michael Bloomberg, made the neighborhood a giant pedestrian zone that attracted people dressed as cartoon characters and superheroes" (Fertig, 2015).

Legal Challenges to Restrictions on Panhandling

Although no one had challenged Mayor LaGuardia's ban on busking, there have been a handful of lawsuits challenging restrictions on panhandling. After a court challenge to the laws against panhandling, the U.S. Court of Appeals for the Second Circuit held that the City of New York could ban panhandling in the subways (*Young v. New York City Transit Authority,* 1990). But three years later the same court held that the First Amendment protects the right to beg on public streets, provided it is not "aggressive begging," meaning that panhandlers try to intimidate passers-by into giving them money (*Loper v. New York City Police Department,* 1993). In *Loper,* the appellate court held that the First Amendment protects begging on public streets because it involves communication "indicating the need for food, shelter, clothing, medical care or transportation" (*Loper v. New York City Police Department,* 1993, p. 704).

Legal Challenges to Restrictions on Busking and Street Performances

With regard to precedents set in other states, street performers successfully relied on public forum doctrine and prevailed in court, establishing their right to perform (*Goldstein v. Town of Nantucket,* 1979), give lectures on bagpipes (*Davenport v. City of Alexandria, Virginia,* 1984), break-dance (*Friedrich v. City of Chicago,* 1985, p. 1149), sing lyrics from the Holy Book of the Sikh religion (*Perry v. Los Angeles Police Department,* 1997, p. 1371), or perform along boardwalks on public beaches (*Dowd v. City of Los Angeles,* 2014). Street performers have also won lawsuits in Waikiki, Hawaii (American Civil Liberties Union, 2001), St. Augustine, Florida (City of St. Augustine, 2003), and San Francisco, California (Matier & Ross, 2004). The U.S. Court of Appeals for the Ninth Circuit struck down the City of Santa Monica's requirement for permits for people who wanted to "travel in unison on any public street, highway, alley, or sidewalk" as specified in Santa Monica's Municipal Code, SMMC §4.68.040(a) (*Santa Monica Food Not Bombs v. City of Santa Monica,* 2006, p. 1053). The U.S. Court of Appeals for the Ninth Circuit also struck down the City of Seattle's ordinance requiring street performers to obtain permits as an unconstitutional prior restraint of speech, but held that the City of Seattle *could* establish "precise performance locations" in order to reduce "territorial disputes" among street performers (*Berger v. City of Seattle,* 2009, p. 1049).

In general, these cases have established that street performers have a First Amendment right to appear in public, and are not required to obtain permits before they perform. Cities *may* rely on time-place-manner restrictions to regulate the *times* and *places* where street performers may engage with the general public, however. In New York City, the Street Performers Advocacy Project has posted a set of guidelines for street performers in which they explain: "You have a right to perform in the subways, on the sidewalks, and in the parks of New York City" (Tanenbaum, Street Performers Advocacy Project & City Lore, 2012).

New York courts have also recognized the value of "place." In 1994 Norbert Bissinger was convicted for taking photographs of people in Times Square without a vendor's license. Prosecutors argued that taking photos of passers-by on the street and selling them to the individuals for $5 was the equivalent of a "shoe shine or sharpening shears" (*People of the State of New York v. Bissinger,* 1994, p. 670). Bissinger countered that he was engaged in "performance art," in which he created a "festive and joyous figurative and literal picture of New York City" for tourists and others, "in contrast to the cold, indifferent, rushing masses that often occupy our streets and heighten a visitor's negative impression of our City." The court held that to deny Bissinger access to Times Square "effectively closes off the only method for defendant here to capture and communicate the idea he assertedly wishes to express – a performance and photographic memorialization, with street audience participation in a bustling street setting." No alternative sites were available because he "could not purchase the atmosphere of a particular Times Square area street corner at any price" (*People of the State of New York v. Bissinger,* 1994, p. 675). The court thus held that the First Amendment protected Bissinger's Times Square photos.

Two decades after Mayor John Lindsay had lifted the ban on busking in New York City, Robert Bery and a group called Artists for Creative Expression on the Sidewalks of New York City filed suit against the City of New York and Mayor Rudolph Giuliani, arguing that the Administrative Code of the City of New York (§20–452(b), "General Vendors Law"), which required the artists to obtain a general vendors license before selling their work in public places, was a violation of their First Amendment rights. The law held that everyone who sold goods or services in a public place that is not a store was required to have a General Vendors license.

The artists charged that the General Vendors license should not apply to them; furthermore, police had harassed or threatened them and had confiscated and sometimes damaged their paintings. Attorneys for New York City argued that the artists were free to exhibit their work in public; they simply could not sell it without a general vendors' license. The U.S. Court of Appeals for the Second Circuit noted that New York City gave out only 853 vendors' licenses at any one time, and the 500 to 5,000-person waiting list "makes [the artists'] prospects of securing a license apparently non-existent" (*Bery v. City of New York,* 1996, p. 697). The Second Circuit continued: "The City apparently looks upon visual art as mere 'merchandise' lacking in communicative concepts or ideas … [but] the sale of protected materials is also protected" (*Bery v. City of New York,* 1996, p. 695). The appellate court explained that "art should be available to the public: "Anyone, not just the wealthy, should be able to view it and to buy it. Artists are part of the 'real' world; they struggle to make a living and interact with their environments. The sale of art in public places conveys these messages" (*Bery v. City of New York,* 1996, p. 697). The Second Circuit thus ruled in favor of the artists, holding that they did not need a general vendor's license to sell their paintings. The law now holds that anyone who sells newspapers, periodicals, books or pamphlets, and anyone who sells artwork including paintings, photographs, prints and sculptures does *not* need a general vendor's license; in other words, if one is selling material that the First Amendment protects, there is no need to get a general vendor's license first. Anyone who sells food must get a license from the Department of Health.

Just as the Second Circuit had ruled in favor of artists selling their paintings on public streets, the Second Circuit also ruled that New York City police could not confiscate buskers' musical instruments; musicians are permitted to use amplifiers if they pay $45 a day for a permit to use amplification, but they are not allowed to use amplification in the subways (*Turley v. City of New York*, 1999).

Does the First Amendment Protect Panhandling?

The Times Square desnudas and the men who dress up as characters from popular culture are proposing a commercial transaction: in exchange for posing for photos with tourists, the tourists reward them with tips. Of course, the First Amendment protects commercial speech (*Virginia State Board of Pharmacy v. Virginia Citizens Consumer Council*, 1976), so the street performers have the right to coax tourists to pose for these photos. Street performers may not cross the line into "aggressive panhandling" and harass tourists for tips, however.

The costumed characters and desnudas in Times Square are panhandling in the sense that they are hoping for "donations" or "tips" from passers-by; in other words, their panhandling is legal, provided that tourists do not feel threatened. Persuading tourists to tip them creates a need for very effective communication as the costumed characters and desnudas interact with potential customers in order to create "feel-good" moments to put these customers in a "tipping mood." In this regard, the costumed characters are just like all other buskers and street performers.

In other jurisdictions across the United States when plaintiffs have challenged anti-panhandling ordinances, federal district courts have struck them down in Grand Junction, Colorado (*Browne v. City of Grand Junction, Colorado*, 2015), Worcester, Massachusetts, (*Thayer v. City of Worcester*, 2015), Lowell, Massachusetts (*McLaughlin v. City of Lowell*, 2015), and Springfield, Illinois (*Norton v. City of Springfield*, 2015). The federal district courts in these cases found the anti-panhandling ordinances unconstitutional because they had singled out a certain kind of speech (asking for money) for special treatment without a compelling reason. Indeed, "dozens of cities have repealed their [anti-panhandling] laws to head off litigation with advocacy groups" (Palazzolo, 2017, p. A3). This change has occurred because, following the Supreme Court's ruling in *Reed v Town of Gilbert* (2015), lower courts have re-categorized restrictions on panhandling as "content-based speech" rather than "content-neutral speech." In other words, if panhandling ordinances are "content-based," they must survive "strict scrutiny;" since very few regulations can pass this legal test, cities across the United States have rescinded their restrictions on panhandling (Palazzolo, 2017, p. A3).

In view of these federal court decisions upholding the First Amendment right to panhandle or solicit tips, the members of the New York City Council looked for guidance to see how city officials in Las Vegas had dealt with this issue. The City of Las Vegas had banned street performers from soliciting tips on Fremont Street. After the U.S. Court of Appeals for the Ninth Circuit ruled that the "Fremont Street Experience unmistakably possessed the characteristics of a traditional public forum" (*American Civil Liberties Union of Nevada v. City of Las Vegas*, 2006, p. 791), however, city officials had to figure out how to rein in scantily clad street performers similar to the desnudas in Times Square. Las Vegas City Councilman Bob Coffin expressed concern about what tourists might run into on Fremont Street: "They don't expect to see women dressed as nuns with their breasts flying everywhere and just their naughtiest parts covered, or codpiece-clad cowboys" ("Las Vegas attempts to tame naked street performers," 2015). Las Vegas now requires street performers on Fremont Street to confine themselves to one of a few dozen 6-foot circles painted on the pavement. The City of Las Vegas administers a daily lottery system that randomly assigns street performers to a 6-foot circle. Performers must change to a different circle every two hours. One purpose of the regulations is to prevent what had occasionally been violent turf wars over busking spots. The privately held company Fremont Street Experience LLC patrols the performance zones, although the City of Las Vegas actually owns the Fremont Street Mall (DeHaven, 2015).

New Regulations = Teal Boxes

After studying the precedent that Las Vegas had set, ultimately the New York City Council passed a "Mascot Law," with the title "Costumed Individuals." Its purpose is to rein in the costumed characters and desnudas. Mayor Bill de Blasio signed the bill, which permitted the New York City Transportation Department to create new rules for Times Square.

Arguing in favor of the new rules, legal scholar Joseph Fan explains that the Mascot Law serves a compelling government interest in "reducing urban congestion for the convenience and safety of its citizens" (Fan, 2015, p. 166). Fan further argues that this regulation would protect the costumed characters and desnudas from hostile police action because everyone would "have the same playbook" (Fan, 2015, p. 167). Fan acknowledges a certain irony regarding the (less visible) economic aspects of the new rules: "The Mascot Law actually legitimizes something illegal: the off-the-books earnings of the mascots" (Fan, 2015, p. 170).

Thus, in June 2016 workers painted teal 8-by-50-foot boxes or "activity zones." As of June 21, 2016, street performers have to stay in the activity zones, whereas passers-by may walk in "pedestrian flow zones." Signs are posted to remind tourists that tips are optional. If street performers break the rules, a police officer can issue them a summons or even arrest them; they can face a $500 fine or even jail time (Fitzsimmons, 2016, p. A16). Some "characters" have strongly objected, saying it is costing them money.

Although some costumed characters are not happy about the new guidelines, "Naked Cowboy" Robert Burck said tips were up: "I love the boxes. I never would have thought I would" (Fitzsimmons, 2016, p. A19).

Discussion

In sum, it is safe to say that there is a First Amendment right to solicit tips, provided the costumed characters or desnudas do not harass tourists or act too aggressively. It is also legal for the desnudas to be topless; if men can be out in public with their chests bare, then so can women.

Although busking was illegal under Mayor Fiorello LaGuardia in the 1930s, buskers and street performers eventually prevailed on Mayor John Lindsay to make their activities legal in the 1970s. Times Square would qualify as a "traditional public forum" subject to time-place-manner restrictions which must be content neutral, but street performers are of course far more concerned with the *time* and *place* regulations as opposed to any issues of *content* because they are not expressing political or religious viewpoints; they are merely posing with tourists for tips. The artists who wished to sell their paintings on city streets also set an important precedent for later street performers because the court held that anyone selling material that the First Amendment protects does *not* need to obtain a general vendor's license first (*Bery v. City of New York,* 1996, p. 697). Just as Norbert Bissinger had a First Amendment right to *take* (and sell) photos of tourists in Times Square (*People of New York v. Bissinger,* 1994, p. 675), the costumed characters have a First Amendment right to *pose* for photos of tourists (and ask for tips) in Times Square.

Not only does the First Amendment protect buskers and street performers, it likewise protects panhandling, provided that it is not "aggressive begging" involving an attempt to intimidate pedestrians into giving the beggar money. As is discussed above, the U.S. Court of the Appeals for the Second Circuit upheld the rights of panhandlers to inform passers-by of their needs "for food, shelter, clothing, medical care or transportation" (*Loper v. New York City Police Department,* 1993, p. 704). More recently, federal courts in Colorado, Massachusetts and Illinois have struck down anti-panhandling ordinances, finding as in *Loper* that panhandling comprises protected speech.

Former Director of the New York Civil Liberties Union Norman Siegel has voiced concern that confining the street performers to the teal boxes could be violating their First Amendment rights: "What the City of New York is creating is a no-free-expression-zone … in the quintessential public space which is Times Square" (Haller, 2016). But Times Square Alliance President Tim Tompkins, who helped draft the new zone regulations, disagrees with Siegel: "There began to be this revelation that this was not artistic expression; it was commercial activity. This is vending on the street in which somebody is selling pictures in a public space where the City has a right to regulate it" (Haller, 2016).

New York City has done well to emulate the City of Las Vegas' policy of assigning street performers to 6-foot painted circles for 2-hour shifts on Fremont Street; this appears to be an effective solution for ensuring that street performers do not obstruct traffic and do not get into turf wars for the best locations. In Times Square, keeping the costumed characters and desnudas in the teal boxes also appears to be working well.

In passing this new law, the New York City Council has found an effective balance between the First Amendment rights of the costumed characters and desnudas and the rights of passers-by to be free from the aggressive panhandling of the Spiderman who punched a police officer, two brawling Statues of Liberty, and Bad Elmo. In this author's opinion, these regulations are in line with Constitutional protections because the regulations are content neutral. The regulations serve a compelling government interest because they balance the need to maintain the free flow of foot traffic and the need to support the street performers' freedom of expression. The regulation is narrowly tailored, and the costumed individuals have sufficient alternative channels of communication. For New Yorkers, the lessons learned are that it *is* possible to protect street performers' First Amendment rights, even if the performers feel a bit boxed in.

References

Adderley v. Florida, 385 U.S. 39 (1966).

American Civil Liberties Union. (2001, December 28). ACLU wins artistic expression lawsuit on behalf of Waikiki street performers. Retrieved February 18, 2016 from https://www.aclu.org/news/aclu-wins-artistic-expression-lawsuit-behalf-waikiki-street-performers

American Civil Liberties Union of Nevada v. City of Las Vegas, 466 F.3d 784 (9th Cir. 2006).

Berger v. City of Seattle, 569 F. 3d 1029 (9th Cir. 2009).

Bery v. City of New York, 97 F.3d 689 (2nd Cir. 1996).

Browne v. City of Grand Junction, Colorado, Civil Action No. 14-CV-00809-CMA-KLM (U.S. District Court for the District of Colorado, filed September 30, 2015).

Campbell, P. J. (1981). *Passing the hat: Street performers in America.* New York, NY: Delacorte Press.

City Lore. (2015). Urban culture. Retrieved February 9, 2016 from http://citylore.org/urban-culture/resources/street-performers/

City of St. Augustine, Florida. (2003, March 24). Draft plan for St. Augustine buskers. Official Web Site of the City of St. Augustine. Retrieved February 18, 2016 from http://www.staugustinegovernment.com/the-city/featured-stories-archive/3_03/ordinance_buskers.cfm

Clarke v. Community for Creative Non-Violence, 468 U.S. 288 (1982).

Cox v. Louisiana 379 U.S. 536 (1965).

Davenport v. City of Alexandria, Virginia, 748 F.2d 208 (4th Cir. 1984).

DeHaven, J. (2015, September 16). Las Vegas okays strict rules for Fremont street performers. *Las Vegas Review Journal.* Retrieved February 7, 2016 from http://www.reviewjournal.com/news/las-vegas/las-vegas-oks-strict-rules-fremont-street-performers

Dowd v. City of Los Angeles, 28 F.Supp.3d 1019 (United States District Court for the Central District of California, filed May 23, 2014).

Fan, J. (2015, Fall). Note and comment: Legal implications of INT 0467-2014, the New York City "Mascot Law" bill: Commentary, legal arguments and loopholes to consider. *Journal of Civil Rights and Economic Development, 28*, 149.

Fertig, B. (2015, August 18). Mayor says he won't tolerate characters harassing people in Times Square. *WNYC.* Retrieved February 9, 2016 from http://www.wnyc.org/story/mayor-times-square-characters-im-not-going-tolerate-it/

Fitzsimmons, E. G. (2016, June 22). Minnie Mouse and Elmo are corralled in Times Square, with mixed feelings. *New York Times,* pp. A1, A16, A19.

Friedrich v. City of Chicago, 619 F. Supp. 1129 (U. S. District Court, Northern District of Illinois, Eastern Division, 1985).

Goldstein v. Town of Nantucket, 477 F. Supp. 606 (D. Mass. 1979).

Greer v. Spock, 424 U.S. 828 (1976).

Haller, V. (2016, August 3). After years of bad behavior, the costumed characters of New York's Times Square no longer roam free. *Los Angeles Times.* Retrieved December 15, 2016 from http://www.latimes.com/nation/la-na-taming-times-square-20160726-snap-story.html

Hawkins, R. (2012). Industry cannot go on without the production of some noise: New York City's street music ban and the sound of work in the New Deal Era. *Journal of Social History, 46*(1), 106–123.

International Society for Krishna Consciousness v. Lee, 505 U.S. 672 (1992).

Las Vegas attempts to tame naked street performers. (2015, July 30). *New York Post.* Retrieved February 6, 2016 from http://nypost.com/2015/07/30/las-vegas-attempts-to-tame-naked-street-performers/

Loper v. New York City Police Department, 999 F. 2d 699 (2nd Cir. 1993).

Matier, P., & Ross, A. (2004, April 7). Bushman of Fisherman's Wharf gets the last ugga-bugga: District Attorney drops stack of complaints over his sidewalk antics. *San Francisco Gate.* Retrieved February 18, 2016 from http://www.sfgate.com/bayarea/matier-ross/article/Bushman-of-Fisherman-s-Wharf-gets-the-last-3313974.php

McCracken, J. (2006). The price and politics of performing. *Urban Action 2006: A Journal of Urban Affairs, 28,* 44–48.

McLaughlin v. City of Lowell, 2015 U.S. Dist. LEXIS 144336 (Dist. Mass. October 23, 2015).

New York task force to suggest new rules for topless Times Square performers. (2015, October 1). *The Guardian.* Retrieved February 9, 2016 from http://www.theguardian.com/us-news/2015/oct/01/new-york-taskforce-topless-times-square-performers

Palazzolo, J. (2017, August 9). Courts overturn panhandling laws: Cities' restrictions on begging are toppling around the United States after a Supreme Court ruling. *Wall Street Journal,* p. A3.

People of the State of New York v. Bissinger, 163 Misc. 2nd 667; 625 N.Y.S. 2nd 823; 1994 N. Y. Misc. LEXIS 609 (Criminal Court of the City of New York, October 24, 1994).

Reed v. Town of Gilbert, Arizona. 135 S. Ct. 2218 (2015).

Santa Monica Food Not Bombs v. City of Santa Monica, 450 F.3d 1022 (9th Cir. 2006).

Tanenbaum, S. J. (1995). *Underground harmonies: Music and politics in the subways of New York.* Ithaca, NY: Cornell University Press.

Tanenbaum, S. J. (2012). Street Performers Advocacy Project & City Lore. Know your rights! A guide for subway musicians and other performers. Retrieved September 30, 2016 from http://citylore.org/urban-culture/resources/street-performers/

Times Square topless ladies. (2015, October 11). *Daily Mail.* Retrieved January 31, 2016 from http://www.dailymail.co.uk/news/article-3268606/Times-Square-topless-ladies-say-refuse-leave-despite-city-officials-desires-confine-performers-activity-zones.html

Turley v. City of New York, 167 F.3d 757 (2nd Cir. 1999).

Tushnet, M. (2012, Winter). Art and the First Amendment. *Columbia Journal of Law & Arts, 35,* 169.

Virginia State Board of Pharmacy v. Virginia Citizens Consumer Council, 425 U.S. 748 (1976).

Ward v. Rock against Racism, 491 U.S. 781 (1989).

Young v. New York City Transit Authority, 903 F.2d 146 (2nd Cir. 1990).

Zack, I. (2015). *Say no to the devil: The life and musical genius of Reverend Gary Davis.* Chicago, IL: University of Chicago Press.

Zick, T. (2006). Speech and spatial tactics. *Texas Law Review, 84,* 581.

CHAPTER EIGHT

Privatopias and Freedom of Expression

Speech Problems in Paradise

DONALD FISHMAN
Boston College

In 2012, Dianne Black placed a political sign supporting a presidential candidate outside her home in St. Paul, Minnesota. Black lived in Carriage Crossing One, a condominium complex in St. Paul's Battle Creek neighborhood. Declared Black: "I ha[d] been diligently working (on) the campaign and thought spending $5 on a yard sign would be a good way to show my support" (O'Connor, 2012). One of her neighbors complained about the sign to the Greystone Management Corporation that oversees the condominium association, and the association then forced Black to remove the sign. Meanwhile, Black remained actively involved in the political process, but she was outraged that she could not express a political point-of-view in the time-honored American way of using a lawn sign (O'Connor, 2012).

Unfortunately, Dianne Black lived in a condominium association, and residents of condominium associations, townhouses, single-family developments, and common-interests communities sign documents when they buy a residence agreeing to various bylaws and covenants. These agreements are binding contracts that can be upheld by law. The Carriage Crossing One Condominium Association has a lengthy list of regulations for its community. Among them is the stipulation that "No 'For Rent,' 'For Lease,' political signs, or advertising signs shall be placed on any part of the property by any person, unless specifically approved by the Board" (O'Connor, 2012). In effect, individuals like Dianne Black surrender their freedom of speech in order to live in their residence. For Black and others, this amounts to giving up their constitutional right "just to find a place to live" (O'Connor, 2012).

Black complained to the Office of the Secretary of State in Minnesota, but its spokesman, John Kavanagh, explained that when you talk about freedom of speech, you are talking about "freedom from government interference in speech." In the case of Dianne Black, "you are talking about a private matter" (Gombossy, 2012).

The restrictions in a condominium complex or a cooperatively owned apartment can be stringent. Even putting a sign in a window is likely to trigger a response from the condominium association. Black's professional management team, the Greystone Group, works with more than 40 associations, and at least 35 have a "no sign in the window" policy. According to Greystone representative Jamie Chapman, such a restriction "helps keep harmony among the neighbors," and the underlying purpose of such a regulation is to avoid "unneighborly dispute" (O'Connor, 2012). Others believe that the purpose of such a restriction, apart from not irritating the neighbors, is to preserve property values (Fleming, 2006, p. 24; Low, 2004, pp. 153–173; McKenzie, 1994, p. 19), and that maintaining property values is why many associations forbid political yard or window signs.

The situation in Minnesota is not unique. Recent developments in the housing market have created a changing pattern of residential living in the United States. "Every year," writes Richard Harris (2013), "more and more Americans are buying into residential developments where many facilities—hallways and foyers, roads and play areas, games rooms, pools, and gates—are held in common" (p. 718). Older neighborhoods in some cities are building condominium complexes and closing off streets to enhance security and reduce traffic (Grant & Mittesteadt, 2004, p. 913). The generic name for these facilities is common-interest communities, or CICs. According to Fleming (2006), "[H]omeowner associations govern more than half of all houses for sale in the fifty largest metropolitan areas" (p. 574). Since the 1960s, CIC facilities have become the predominant form of new housing construction (McKenzie, 2011, p. 26). These housing arrangements provide affordable, hassle-free ownership; buyers keep the title-deed, but they share the responsibilities for maintenance. Over sixty-three million people, or one in five U.S. residents, now live in a homeowner association, condominium complex, townhouse developments, housing cooperative, or some other form of planned community (Harris, 2013, p. 717; McKenzie, 2011, p. 2).

Moreover, these home or condominium associations have extensive powers. Observes McCabe (2011): "Homeowner associations hold elections, provide services, tax residents, and regulate behavior within their jurisdiction" (p. 535). In these private housing communities, basic First Amendment rights have been restricted in ways that the Founding Fathers in 1787 never would have envisioned. Individuals in these housing developments frequently are compelled to trade their constitutional rights for the privilege of living in a particular community.

One of the prominent names for such innovative housing arrangements is "privatopia." The term embraces a whole range of private communities that have emerged in the suburbs and in the central cities, especially with the increasing popularity of back-to-the-city condominium conversions. It is a catch-all phrase that reflects two strands of thought: Privatization and planning. McKenzie (1994) views privatopias as a growing "*ad hoc* form of privatization that carries with it significant social and political consequences" (p. 26).

Concurrently, many cities have discovered that privatopias are economically beneficial: The residents pay property taxes but do not utilize a full share of municipal services. "The developments," writes McKenzie (1994), "take over many municipal functions for those who can pay the price" (p. 21). CICs become mini-governments, or quasi-governmental entities, balancing the rights of their members with the demands of the outside world. Yet, in terms of a traditional conception of freedom of speech, there is a dark side to these privatopias that raises serious constitutional and legal issues: Governing associations frequently go beyond what a municipality would do to a private, property owner in prohibiting speech activities.

The purpose of this chapter is three-fold: First, to discuss the emergence of privatopias as a new housing paradigm. Second, to examine several key legal cases where traditional freedom of speech rights have been restricted within privatopias. Third, to explore the prospects for strengthening First Amendment protections in privatopias using government-based regulations to remove restrictions that impair communicative expression.

Privatopias as a New Housing Paradigm

A "privatopia" refers to a planned or gated community in which a homeowner or condominium association establishes rules pertaining to the appearance of the properties and the behavior of the residents. These privatopias are spatially defined residential communities with shared amenities and regulatory controls imposed on communicative expression. Under the over-arching goal of promoting civic harmony and social peace, an association attempts to rein in individuality by requiring standards of behavior that are formal, objective, easily understandable, and can be enacted as simple clauses in the initial sales contract of a residential unit (Blakely & Snyder, 1999; Grant & Mittesteadt, 2004; McKenzie, 1994). Frequently formed as a special district in the absence of an official municipality, such associations also maintain and operate their own infrastructure, including contracting privately for water, sewer, trash, and fire services. Governance is delegated to an elected board which often subcontracts to a professionally-operated management team to run the activities of the housing complex. In this arrangement, owners surrender some control over their homes and, in return, they gain similarly limited control over their neighbors' homes.

Privatopias are the fastest growing segment of the housing market in the United States. The numbers are staggering. In 1970, 2.1 million people lived in privatopias; by 2008, that number rose to 59.5 million. By 2012, it was estimated that 63 million Americans now live in 23 million households in these associations that share the responsibilities of neighborhood governance through voluntary memberships in 323,000 community associations (Harris, 2013, pp. 717–718; O'Connor, 2012). From Florida to California, from Texas to Maine, more and more people are entering gated communities or various forms of homeowner or condominium associations.

These associations, whether co-op apartments in New York City or fenced-in communities called "plantations" in Georgia, have led Harris (2013) to interpret the increasing desire for these new housing arrangements as "The American Dream Rebranded" (p. 717). The "home-as-castle" image has given way to "affinity living" within a homeowner's association. More and more Americans are buying into such residential developments. "Market demand," write Grant and Mittesteadt (2004) "results in a large proportion of new communities in the U.S. being gated" (p. 914). In inner cities, barricaded streets and restricted entry roads frequently are used to segregate the residential spaces (Grant & Mittesteadt, 2004, p. 923). Unfortunately, occupants of privatopias do not necessarily read the fine print in the sales agreements in what is often a lengthy document called "Covenants, Conditions & Restrictions." As songwriter Tom Waits observed, "[T]he large print giveth and the small print taketh away" (quoted in Harris, 2013, p. 717). But the freestanding, detached home as a symbol of unfettered freedom and a bastion of liberty is slowly receding in importance.

The particular forms of the housing-condo facilities are not central to the focus of this chapter, but these forms often explain, if not predict, the level of divisiveness within residential communities. Approximately, 55 percent of privatopias are condominium associations; the remaining 45 percent are almost all homeowners associations (Kennedy, 1995, p. 765). Nonetheless, there are important differences among the various condo-like arrangements. Cooperative units vote-in members while condominium complexes accept a new member at the discretion of the owner of the unit. In a cooperative arrangement, a board controls the sale and leases of a residential unit. For instance, the Dakota, the legendary New York City apartment building, is governed by a board that has rejected would-be buyers such as Billy Joel, Cher, Melanie Griffith, and Antonio Banderas (Lattman & Haughney, 2011). In another prominent case, Peter Lawford, a Hollywood actor, a Rat-Pack member, and the brother-in-law of President John Kennedy, had his application turned down for a sixteen-room apartment at 117 East 72d street in

New York City on the grounds that he was an actor and a Democrat. The city's Commission on Human Rights declined to intervene in the Lawford case because the rejection did not occur on the basis of suspect categories such as race, creed, religion, or nationality (Bower, 1992, p. 206; "Peter Lawfords Barred from Eastside Co-op," 1964). It is not surprising, therefore, given the closely monitored application process, that co-op units are more homogeneous and more likely to seek consensus while condominium associations have been more heterogeneous and more likely to be the site of friction (Bower, 1992, p. 219; McKenzie, 1994, pp. 19–21).

Equally important, this new paradigm of shared property ownership and private government activity carries out many responsibilities formerly provided exclusively by local government. Contrary to expectations, many municipalities have welcomed these developments. Local governments view privatopias as cash cows—their residents pay property taxes but do not use the full range of municipal services. Thus, it is not surprising that local governments have been eager to provide construction permits for this kind of residential living. Moreover, local governments do not view the governance boards of privatopias as competitors or eventual replacements for their functions. Instead, homeowner associations have become the most common and fastest growing units of local governance in the United States. One director of a homeowner association refers to them as "banana republics" (Lloyd, 2009) while Gott (2013) views them as "Ticky Tacky Little Governments." They often function as mini-governments or quasi-government entities creating rules and regulations independent from governmental oversight.

Privatopias have been an innovative solution to a myriad of difficult issues related to residential property. These housing associations have prospered because they solve collective action problems (shared landscaping; shared swimming pools; shared snow plowing; shared recreational facilities; and often shared but restricted golfing areas), and they fulfill the desires of their residents to live in residential harmony. But in creating the perfect community, privatopias frequently go too far and infringe upon the fundamental liberties of property owners. Privatized local government does not, or should not, translate into "privatized constitutional rights," but the courts generally have been deferential to privatopias.

Legal Issues Related to Privatopias

There have been several federal, state, and local cases that have raised issues regarding the scope of expressive communication in a privatopia. Most privatopias have non-negotiable contracts that require would-be buyers to sign away their First Amendment rights. These privatopias regard themselves as business entities, not fledgling democracies. Owners and managers of these privatopias insist that they are not legally obligated to provide a venue for the exercise of freedom of expression. In their earliest stages, these privatopias have been treated as private parties, not state actors (Kennedy, 1995, p. 761). By and large, the legal cases illustrate the tensions between traditional values of freedom of speech and the belief that "the infringement on one's own freedom is a small price to pay for the protection from the potential misdeeds of one's neighbors" (McKenzie, 1994, p. 14).

At bottom, people who inhabit and interact in these residential spaces face demands that run counter to traditional First Amendment jurisprudence. Most importantly, as the number of people living in these communities continues to increase, more and more conflicts related to defining the rules and scope of governance is likely to occur. Five recent legal cases provide insight into the various tensions surrounding the wide-ranging efforts to impair freedom of expression.

(1) City of Ladue v. Gilleo (1994)

In *City of Ladue v. Gilleo* (1994), the Supreme Court addressed the issue of a city-wide ban on yard signs designed to minimize visual clutter within a municipality. In 1990, the plaintiff, Margaret P. Gilleo placed a yard sign on her front lawn expressing her opposition to the Persian Gulf War. The sign stated "Say No to War in the Persian Gulf, Call Congress Now" (p. 45). Gilleo's first sign disappeared, and her second sign was knocked down and damaged. When Gilleo reported these incidents to the police, she was informed that such yard signs were prohibited in the city of Ladue. Ladue is a modest suburb of St. Louis, and it is zoned for both commercial and industrial uses, which means that a concern with signage clutter was a legitimate issue.

Ladue's broad-based ban was designed to reduce visual clutter, allowing exceptions for ten categories of signs, which included residential markers, home sale signs, and commercial signs in properly zoned areas. Gilleo, in turn, then filed a lawsuit in federal court against the Mayor and the members of the city council. The District Court found the city's ordinance unconstitutional. The Court of Appeals affirmed the decision, holding that Ladue's ordinance was a "content-based" regulation.

After the District Court's decision, Gilleo created a 8.5 by 11-inch sign and placed it in the second story of her home window. The new sign said: "For Peace in the Gulf." The Ladue City Council responded by passing a new ordinance prohibiting signs, specifically identifying window signs as a banned item. Gilleo sued again, and the Court of Appeals again affirmed her position. The City of LaDue brought the case to the Supreme Court.

Justice Stevens wrote the opinion for a unanimous Court. Stevens acknowledged that signs are a form of expression protected by the Free Speech Clause, but he contended that signs faced distinctive problems because they are subjected to the police powers of a municipality (p. 48). Unlike oral speech, signs posed special problems because they take up space, may obstruct views, distract motorists, displace alternative uses of the land, and create other problems that warrant legitimate "calls for regulation" (p. 48). Stevens noted that the Court in *Metromedia v. San Diego* (454 U.S. 511–512) had employed the standard of visual clutter to prohibit the use of off-site commercial billboards, but that several members of that Court had recognized that billboards were "an effective medium of communication," even for "non-commercial messages" (p. 50).

Stevens observed that the City of Ladue justified its ban on yard signs upon two major arguments: (1) The City's restrictions were content-neutral; and (2) The City's concern with avoiding visual clutter justified "the comprehensiveness of the sign prohibition" (p. 51). In fact, the City of Ladue argued that the mix of prohibitions and exemptions in its ordinance reflected "legitimate differences among the side effects of various kinds of signs" (p. 52). These differences, the City insisted, were not related to the contents of the message nor whether the City approved or disapproved of the viewpoint. Stevens, however, was not persuaded by the City's rationale and he expressed the belief that the City's ban actually "prohibits too much speech" (p. 52).

Moving away from the issue of the "undesirable secondary effects" of visual clutter, Stevens then focused on what he believed to be the threshold question raised by the case: Gilleo asserts "a constitutional right to display an antiwar sign at her own home" (p. 53). The subsidiary questions were whether the City of Ladue may legitimately prohibit her from displaying such a sign, and whether it is proper to permit certain signs while creating a near-total ban on other signage.

Stevens addressed the first question about the rights of a speaker with an eye to traditional First Amendment jurisprudence. He worried that allowing a total ban on yard signs would foreclose a "venerable means of communication that is both unique and important" (p. 54). He contended that residential signs play an important part in political discourse—allowing an individual to react quickly, personally, and creatively to political concerns. While foreclosing an entire medium may be advisable to formulate a content-neutral approach, the process, in turn, unfortunately eliminates a common means of speaking and "such measures can suppress too much speech" (p. 55). Stevens, therefore, rejected the City of LaDue's contention that its ordinance was merely a "time, place, and manner" regulation because residents still were free to convey the message by other means, such as hand-held signs, flyers, bumper stickers, or newspaper advertisements. Instead, Stevens argued that he was unpersuaded that an adequate substitute existed for displaying a sign on one's own residence. Wrote Stevens: "Precisely because of their location, such signs provide information about the identity of the 'speaker'" (p. 56).

Stevens suggested that a sign such as "Peace in the Gulf" on the front lawn of a retired general or decorated war hero provokes a very different reaction than the same sign in the window of a ten-year old child or a bumper sticker on a passing automobile (p. 56). The argument set forth is that the identity of a speaker is critical, and signs at a particular residence forge a special relationship between a speaker's identity and a location. Moreover, Stevens insisted that residential signs are cheap and convenient, and that they can be displayed by people of modest means or limited mobility. Stevens concluded that a concern for visual clutter was real, but that the self-interest of the residents to keep up their own property "diminishes the danger of the 'unlimited' proliferation of residential signs that concerns the City of Ladue" (p. 58). Steven's theoretical approach was anchored in the rights of a speaker, a speaker's identity, and the speaker's use of his or her home as the background for the message. Instead of avoiding the neighbors and reducing the likelihood of bothering them, the Court assumed that the speaker deliberately wanted to reach his or her neighbors. Stevens believed that there was a corollary right to do so within the sphere of the First Amendment.

(2) Twin Rivers Home Owners Association (2007)

In the *Committee for a Better Twin Rivers v. Twin Rivers Homeowners Association* (2007), the Supreme Court of the State of New Jersey addressed whether the rules and regulations enacted by a homeowners association governing (A) the posting of signs, (B) use of the community room, and (C) access to an association's newsletter violated that state's constitutional guarantees of freedom of expression. The trial court held in favor of the homeowners association while the New Jersey Appellate division later reversed that decision. Thus, the Supreme Court of New Jersey was forced to examine directly the type of friction that arises between rules and regulations of a homeowner's association and the traditional rights of freedom of expression.

Twin Rivers is a planned development community of privately owned condominium duplexes, townhouses, single-family homes, apartments, and commercial buildings located in East Windsor, New Jersey. The community covers one square mile, and it had approximately 10,000 residents. A lawsuit was brought by three members of the community seeking to invalidate various rules and regulations enacted by the Board. The Association's response to the court stressed four major points: (1) That the court should follow other jurisdictions that have refused to impose constitutional obligations on private homeowner associations; (2) That the homeowner association does not allow the public to use its property; (3) That the Association's relationship with its members is a contractual one as set forth in a reasonable and restrictive covenant; and (4) That the members participate fully in the decision-making process of the Association. The over-arching question before the New Jersey Supreme Court was whether in the context of a private and contractually-based community, that the Association's rules and regulations were constrained by the constitutional rights of the plaintiffs.

The New Jersey Supreme Court's analysis relied heavily on the precedents established in the U.S. Supreme Court cases on company towns and shopping malls. In order to answer the question about the right of plaintiffs to post signs, use the community room, and access the Association's newsletter, the New Jersey Supreme Court examined *Marsh v. Alabama* (326 U.S. 501, 1946). In *Marsh*, a private corporation controlled all aspects of Chickasaw, Alabama, a company town near Mobile, Alabama. Among its prohibitions, the company refused to allow the solicitation and distribution of religious literature. Defendant Marsh was a Jehovah's Witness who was charged with distributing religious literature on company-owned land that otherwise was open to the public.

In a 5-3 decision, the U.S. Supreme Court balanced the rights of the property owners against the rights of citizens to enjoy freedom of speech, concluding that the property rights of a private entity are not sufficient to overcome a citizen's fundamental right to speak (*Marsh*, 509). The *Marsh* Court emphasized that the more an owner opens the property to the public, the more those property rights are circumscribed by the statutory and constitutional rights of those who are invited in. The Court concluded that in the context of this particular company town, there was enough activity to say that the property owner's activity constituted state action and thus violated the First Amendment (*Marsh*, 508–509). Twenty-two years later, the Supreme Court reaffirmed its holding in *Marsh* in a privately-owned shopping mall case, *Amalgamated Food Employees Union 590 v. Logan Valley Plaza* (391 U.S. 308, 1968), but the Court later retreated from this expansive position in *Pruneyard Shopping Center v. Robbins* (447 U.S. 74, 1980), concluding that the First Amendment does not grant broad communicative expression rights in privately owned shopping centers.

The state of New Jersey also had its own controlling precedent in *Princeton University v. Schmid* (1982). In *Schmid*, the appellee, who was not a Princeton University student, was arrested for criminal trespass while distributing political materials on campus without first having obtained permission from Princeton as required by university regulations. The defendant claimed that his arrest was unconstitutional because both the First Amendment and Article I of the New Jersey Constitution protected the distribution of political materials.

Princeton University, on the other hand, contended that as a private university the activities on its private property were not subject to the strictures of the federal or state constitution. In *Schmid*, the New Jersey Court formulated a test to balance the rights of property holders against the speech rights of the defendant. The test required the court to consider three elements: (a) The nature, purposes, and primary use of such private property (b) The extent and nature of the public's invitation to the property and (c) The purpose of the expressive activity in relationship to both the private and public use of the property. Using this test, the court concluded that Princeton violated the defendant's "constitutional right of speech and assembly" (*Schmid*, 1982). Most importantly, the *Schmid* test became a key-balancing rubric as New Jersey courts faced issues of private property infused with a public or quasi-public function.

Returning to the issue presented by the Twin Rivers Homeowners Association, the New Jersey State Supreme Court noted that most jurisdictions required "'state action' as a precondition to imposing constitutional obligations on private property owners." Instead, the court asserted that the three-pronged *Schmid* test was the appropriate standard to apply to the Twin Rivers situation. Using the first prong of the test, the court concluded that the nature and purpose of the Twin Rivers community was to serve as a private "common-interest community" that offers services to its residents. On the second prong, the court observed that the Association does not invite the public to use the property. This factor, too, did not support the arguments that the plaintiff's constitutional rights had been compromised. On the third prong, the court held that the plaintiffs' right to post signs, use the community room and access to the community newspaper were not unreasonably compromised "by the regulations of the association."

Above all, the New Jersey court's reasoning emphasized the contractual nature of the community. While posting signs, free use of a meeting room, and access to a newspaper involved politically-oriented speech, the court believed that the relationship between the homeowners association and the plaintiffs was primarily a contractual one, "formalized in a reasonable covenant that appear in all deeds." The court held that the "minor restrictions on the plaintiff's expressive activities are not unreasonable or oppressive, and the Association is not acting as a municipality." In the end, the court overturned the decision of the Appellate Division, and it reinstated the trial court verdict that favored the homeowners association.

(3) Mazdabrook Commons Homeowners Association v. Khan (2012)

In *Mazdabrook Commons Homeowners Association v. Khan* (2012), the New Jersey Supreme Court faced the issue of whether a homeowners association can prohibit residents from posting political signs in the windows of their own homes. In the earlier *Twin Rivers* case (2007), homeowners were allowed to place signs in their windows and in the flowerbeds adjacent to their homes. In *Mazdabrook*, the court faced a similar issue but a very different set of facts.

Wasim Khan lived in Mazdabrook Commons, a residential community in the town of Parsippany, New Jersey. Mazdabrook Commons is a planned community of 194 townhouses in an area that is ranked in one list as one of the top twenty most desirable locations in the United States. In 2005, Khan ran for the Parsippany Town Council, and he posted two signs in support of his campaign in his house—one inside the window of his townhouse and the other inside his door. Mazdabrook Commons notified Khan that the signs violated the association's rules and ordered him to remove them. Mazdabrook Commons' policies prohibited all expressive activity, including political signs except for "For Sale" signs.

The New Jersey Supreme Court noted that Mazdabrook Commons was an independent common-interest community in which individual owners agree to certain common restrictions for the benefit of the entire group. The court focused its analysis in the case on a three-factor test developed earlier in *State v. Schmid* (1982). The court found that the first factor in the *Schmid* test weighed in favor of the homeowners association. First, Mazdabrook is a common-interest community, where owners accede to certain restrictions in the documents that they initially sign. The court held that the primary or normal use of the property is "residential." The second factor of the *Schmid* test also favored the homeowners association: The public is not routinely invited to use the property. Although the public is not barred from Mazdabrook, there is no ongoing invitation to use the streets or shops in the community. Moreover, the plaintiff in the case, Wasim Khan, was not an outside visitor or invited guest. At this point, however, the court did an interesting turn with the first two *Schmid* factors and altered the viewpoint of the case: "From the perspective of the individual homeowner, the first two *Schmid* factors do not favor near-absolute limits on placing a political sign inside one's own home."

The court then turned to the third *Schmid* factor, which became the key element in the case. The court analyzed the purpose of the expressive activity on the property in relation to its private and public uses. The court held that the speech in question was political speech, which it stipulated was at the core of our constitutional scheme for freedom of expression. The court noted that the restriction supported by the association prevented Khan from even promoting his own candidacy for the city council. Quoting from the Supreme Court decision in *City of Ladue v. Gilleo* (1994), the court noted that residential signs are cheap, convenient and important precisely "because of their location," and that they connect the message directly to the speaker and thus the location is intertwined with the words on display.

In *Mazdabrook*, the court also addressed the question about whether Khan actually had waived his freedom of speech in signing the documents to join the common-interest community. At issue was the assertion that Khan had bought his residence with the full knowledge of the restrictions against posting political signs. The court defined a waiver as the "intentional relinquishment or abandonment of a known right or privilege." But the court was doubtful that Khan fully understood the rights he was allegedly relinquishing in the original signing documents to purchase his property: "Can fundamental constitutional rights be properly waived by including waiver language in the midst of a more than fifty-page, single-spaced document?" The court doubted that this constituted a "knowing" and "intelligent" way to waive fundamental constitutional rights. The court instead concluded that because the restriction is inherently "unreasonable" and because it violated the state constitution, that "the covenant that it memorializes is unenforceable."

(4) Bloch v. Frishholz (2009)

In *Bloch v. Frishholz* (2009), the Seventh Circuit addressed a regulation that prohibited the placement of a Mezuzah on the doorpost of three condominium units in a large Chicago building. The Bloch family and its adult children had lived for thirty years in Shoreline Towers, a 378-unit condominium complex in Chicago. In 2001, the Blochs were confronted with a new "hallway rule" that prohibited the placement of "mats, shoes, carts, and objects of any sorts" outside the entrance to their condominiums. The rule later was applied to religious symbols, such as a Mezuzah, a metal object containing a small scroll of parchment inscribed with passages from the Torah, the holiest text in the Jewish religion.

While facially neutral, the hallway rule had been ignored for its first three years. But beginning in 2004, the condominium association began to renovate and paint the hallways and to enforce the hallway rule as an absolute ban. The Blochs, who removed the Mezuzah affixed to their door during the renovations, restored it when the work was done. Without notice to the Blochs, the condominium association began to apply the hallway rule to the Mezuzahs. The condominium association confiscated their Mezuzahs along with the crucifixes, wreaths, Christmas ornaments, political posters, and Chicago Bears pennants affixed to the doors of the other residential units in the complex.

The Blochs voiced their concerns to the condominium association. They provided information explaining the religious significance of the Mezuzah. However, Edward Frishholz, president of the condominium association, rigorously enforced the hallway rule, telling the Blochs if they did not like the rule, they should leave the condominium complex.

The Mezuzah controversy reached a crisis point during the funeral of Marvin Bloch, the patriarch of the family. The Blochs requested that the Mezuzah be left up during the seven days of Shiva, the Jewish mourning period. Frishholz agreed to allow the Mezuzah to remain up during this mourning period along with permitting a coat rack and card table to be placed in the hallway for visitors. Upon returning from the burial of Marvin Bloch, however, the family found that the Mezuzah had been removed again, even though guests, including their rabbi, were gathering at their apartment for a final get-together. On September 16, 2005, the Blochs filed a lawsuit. The City of Chicago later amended its housing code to proscribe applying restrictions in condominium and rental properties to religious symbols on the doorposts. So, too, did the state of Illinois.

But the Blochs continued their lawsuit on the basis of religious post-occupancy discrimination. They contended that the reinterpretation of the hallway rule in 2004 was based on discrimination. At issue was whether the Federal Housing Act could provide relief for a case of post-occupancy discrimination. In a restricted ruling, the Seventh Circuit rendered a decision in favor of the Blochs. The court held that it was unlawful to coerce, intimidate, threaten, or interfere with a person and it prevented a plaintiff from having to "vacate the premises" as had been suggested to the Bloch family.

(5) Flag Flying Controversies

One of the most emotional issues that privatopias encounter is whether or not there should be restrictions on flying an American flag. There is a ripple effect associated with permitting flag flying. Once the door is open to displaying an American flag, this allows other types of flags to be flown such as the British, French, Chinese, or Israeli flags, including flags that have become a political symbol for a social movement, such as the Gay Pride Flag. The emotional dimensions of this issue were vividly displayed in *Gerber v. Longboat Harbour North Condominium Association* (1989). Gerber, an Air Force veteran, attempted to fly the flag outside of the times designated by his condominium association. Gerber sued to remove the time restrictions, and a federal district court found in his favor. The District Court Judge Elizabeth A. Kovachevich wrote:

> It is a curious ordering of values, and a questionable jurisprudence, which forbids a man from displaying the symbol of his country while staunchly defending the rights of others to deface, desecrate, and destroy that same symbol. Had Mr. Gerber chosen to burn his flag rather than display it in a dignified manner, public-spirited lawyers would have appeared to help him protect his constitutional right to burn Old Glory. But to proudly display the United States Flag, Mr. Gerber was forced to commence a federal lawsuit at his own expense.

In *Gerber*, the district court ruled in favor of the plaintiff. The judge held that the condominium's overall activities were sufficiently broad to amount to state action. The court also ruled that the state was obligated to protect the plaintiff and not deprive him of his rights, privileges, and immunities secured by the First and Fourteen Amendments. But as Judge Marvin Aspen in *Goldberg v. 400 East Ohio Condominium Association* (1998) later observed, "[O]ld-fashioned patriotism, rather than legal reasoning, is the source of the *Gerber* opinion's persuasive force" (Goldberg, 1998).

In 2004, George W. Bush signed a bill preventing privatopias from restricting the display of the United States flag by individual unit owners in a homeowners association. Sponsored by Rep. Roscoe Bartlett (R-Md.), the bill passed unanimously in the House and the Senate. "The Freedom to Display the American Flag Act of 2005" stated that the purpose of the measure was "to protect our citizens' right to express their patriotism here at home without burdensome restrictions." The Act makes it impossible for a homeowners association to prohibit residents from flying the American flag. Representative Bartlett stated that "the goal was to find a compromise between the right of individuals to display the flag and the right of homeowners associations 'to maintain the value of their properties'" ("Flying the U.S. Flag," 2009). Several states have enacted similar legislation to protect flag displays and related symbols on condominium property, widely referred to as "flag encouragement laws" (Craig, 2007, pp. 1–12; "Flying the U.S. Flag," 2009).

Privatopias and Government-Based Regulations

Privatopias have not been speech-protective agencies; instead, they have been harmony-maximizing entities. Privatopias fulfill the dream of home or condominium ownership, minimal interference from local government, private services, and the promise of annoyance-reduced interactions with neighbors. In the privatopias, members voluntarily consent to a private, contractual relationship when they move in, and, in theory, they have a voice and are able to participate in how the regulations of the community are promulgated. In actuality, there is little supervision of the abuses of associational boards. As Bower (1992) contends, legal regulations on "disclosure, notice, and voting requirements" create the belief that if the contracting process is regulated, then "substantive policing of the terms of the bargain is unnecessary" (p. 228). This is a strong presumption in favor of the practices of the boards of privatopias. Understandably, courts have been reluctant to interfere with a contractual relationship between two private parties unless the particular provisions violate public policy. The courts have assumed a "knowledgeable consumer" acting voluntarily who requires little, if any, oversight (Bower, 1992, p. 229). There is, however, little evidence that most purchasers read or understand the contracts that they initially are signing (Bower, 1992, pp. 246–247).

By and large, these homeowner and condominium associations have operated beneath the radar screen of public attention. But one in five Americans now resides in a privatopia of one form or another. Private parties now exercise powers over citizens that once were only handled by state entities, which, in turn, raises questions about the treatment of this recently created "privatized" arrangement (McCabe, 2011, p. 539).

The "private" nature of privatopias hinges on a number of assumptions: (1) That residents voluntarily consent to the agreements (2) That there are signed contracts to legitimize these agreements (3) That services once offered by the state are now being replaced by private companies and (4) That residents participate and are able to amend the rules and regulations. But these premises are not immunized from legal inquiries. An association's rules can be subjected to challenges for not being narrowly tailored, not being content neutral, being vague or violating a fundamental right that a resident should not be capable of surrendering under either a state or federal constitution.

Franzese and Siegel (2008) argue that association boards that govern these privatopias need a new paradigm. Among the problems they identify are that many residents are not able to get the attention of their board or manager; boards use their powers to punish owners with opposing views; opposing the board on a particular issue requires extreme efforts and costly litigation; and boards often conduct sham elections to perpetuate their own power (pp. 762–764).

Furthermore, home and condominium purchasers do not realize the extent that association rules will govern their lives. Nor do they realize that the post-purchase relationship between a resident and an association offers few safeguards or that most of their rights may have been waived in signing the initial document. As Bower (1992) writes, most state legislation contains "vague or undemanding standards" for governing boards of privatopias (p. 231).

One of the issues that the skyrocketing growth of privatopias raises is whether the restrictions on political speech triggers state action. The question is: Does the enforcement of private agreements contained in a declaration of a condominium or homeowner contract implicate state action and bring heretofore private conduct within the sphere of the First and Fourteenth Amendment? Many believe that judicial enforcement of private covenants abridging protected speech does not constitute state action (Rahe, 2002, pp. 526–546). That has been the dominant view in the industry. Callies, Franzese, and Guth (2003) observed that the lower courts "for the most part" have resisted applying the state actor doctrine to privatopias (p. 189).

Others believe that the application of First Amendment protection to speech in privatopias is grounded in state action (Alexander, 2012, p. 561; Askin, 1998, pp. 949–960; French, 1992, pp. 350–352; Gott, 2013, p. 201; Pieciak, 2009, p. 174; Rosenberry, 1984, pp. 25–27). They contend that the nature of speech rights at risk in these new housing arrangements is so central to democracy that it warrants a serious re-examination, especially since one in five U.S. residents now live in a privatopia. Moreover, these observers believe that privatopias constitute a form of "self-enslavement," which "is fundamentally repugnant to a free society" and that the civil liberties in question "cannot be bargained away" (Bower, 1992, p. 254). There also may be a distinction needed, such as strict scrutiny for the initial contract and intermediate scrutiny for amendments to the contracts that are amended or changed.

One answer is to apply a First Amendment-type test to homeowner and condominium associations, independent of whether they are deemed state actors. The New Jersey Supreme Court did that in the 2012 *Mazdabrook* case, and the decision highlighted the fact that privatopias, due to their size and importance, now play a critical role in the civic life of our country. That court left undefined the precise dimensions of the standards involved in a constitutional remedy, but it did provide an approach to tackle the problems that millions of people now living in common-interest communities encounter.

Conclusion

The literature on privatopias offers a polarized view of free speech issues involved in these residential housing complexes. One image is that these housing arrangements constitute an idealized form of living arrangements that by 2016 has become firmly entrenched in society. These residential communities located in the inner city, the outer rings of the city, and the suburbs, have become the fastest growing form of new housing construction. Residents of such communities sign agreements that regulate the architectural design of their houses, the size of their doors and windows, the allowable colors of their residences, and place limits on their communicative expression. All of these restrictions are designed to enhance harmonious living conditions.

The other image is that privatopias are disasters waiting-to-happen, and that they impair basic civil liberties that are critical to the operations of a democracy. Writes Suarez (2006): "The loss of the speech right is one of the most severe constitutional deprivations among free people, yet the courts have limited ability to protect private CIC residents and non-residents against it" (p. 741). In this view, privatopias represent a privatization of constitutional protections that ultimately hurts the dynamics of the interactions among residents. This view holds that the "voluntariness" of these agreements is a misnomer, and that the agreements actually are unread and unreadable documents signed under duress. To these observers, the alleged "reasonableness" of some of the provisions in these contracts should be invalidated in court (Suarez, 2006, p. 759).

When privatopias were a small segment of the housing market, the public at large could easily ignore them. Now that over sixty million people live within their structures, the consequences of this residential living arrangement for political and social life in the United States are enormous. The explosiveness of their growth and the popularity of these housing arrangements can no longer be overlooked. Kennedy (1995) argues that left unchecked, privatopias constitute a segregated dimension in American society that will accelerate societal disintegration and disengagement of the wealthier and more skilled members of society from civic issues (p. 777). Former Secretary of Labor Robert Reich (1991) called privatopias the "secession of the successful" transforming the word "community" from neighbors who help one another to economic enclaves of like-minded people disengaged from civic life (p. 41).

Understandably, courts have been apprehensive about imposing regulations upon the operations of such housing arrangements, absent the existence of statutory reforms by legislatures. But such appeals for statutory solutions may be "feigned grief" on the part of the courts, which know that some judicial action is necessary and that legislative reform is not easily forthcoming. These privatopias, their function and status in law, deserve to be re-assessed: A private actor that exercises traditional government authority may actually become a state actor who has entangled the private-public functions of residential housing in a new and challenging way.

The changing nature of privatopias, including the increasingly attractive back-to-the-city condominium conversions, requires swift and flexible judicial action. Due to the rapid growth and continued popularity of privatopias, a First Amendment-type test and the application of the state actor doctrine may be necessary to remedy the existing abuses. If civil liberties are to continue to play their traditional role in American society, the word "community" cannot serve as a pretextual term for "uniformity" and "harmony." That is not the central vision that the First Amendment was designed to promote.

References

Alexander, S. L. (2012). Chapter 383: Expanding freedom of expression for California renters. *McGeorge Law Review, 43*, 553–561.

Amalgamated Food Employees Union 590 v. Logan Valley Plaza, 391 U.S. 308 (1968).

Askin, F. (1998). Free speech, private space, and the Constitution. *Rutgers Law Journal, 29*, 947–961.

Blakely, E. J., & Snyder, M. G. (1999). *Fortress America: Gated communities in the United States.* Washington, DC: Brookings Institution Press.

Bower, T. (1992). Communities within the community: Consent, constitutionalism, and other failures of legal theory in residential associations. *Journal of Land Use and Environmental Law, 7*, 203–273.

Callies, D. L., Franzese, P. A., & Guth, H. K. (2003). *Ramapo*: Looking forward: Gated communities, covenants, and concerns. *The Urban Lawyer, 35*(1), 177–202.

City of Ladue, et al. v. Margaret P. Gilleo, 512 U.S. 43 (1994).

Committee for a Better Twin Rivers v. Twin Rivers Homeowners Association. (2007). Retrieved from http://caselaw.findlaw.com/nj-supreme-court/1363642.html

Craig, B. (2007). Construction and constitutionality of the Freedom to Display the American Flag Act. *Real Estate Law Journal, 36*, 1–12.

Dopplick, R. (2009, January 28). Flying the U.S. Flag: Freedom of speech and private contracts. Retrieved from http://www.inside/justice.com/intl/2009/01/28/u.s._flag_restrictions_laws

Fleming, B. J. (2006). Regulation of political signs in private homeowner associations: A new approach. *Vanderbilt Law Review, 56*, 571–607.

Franzese, P. A., & Siegel, S. (2008). The Twin Rivers case: Of homeowners associations, free speech rights, and privatized mini-governments. *Rutgers Journal of Law and Public Policy, 5*(4), 729–768.

French, S. F. (1992). The constitution of a private residential government should include a bill of rights. *Wake Forest Law Review, 27*, 345–352.

Gerber v. Longboat Harbour North Condominium, 724 F. Supp., 884 M.D. Fla, (1989). Retrieved from http://law.justia.com/cases/federal district-courts/F. Supp./724/884/1468151

Goldberg v. 400 East Ohio Condominium Association, 12 F. Supp. 2d 820 (1998).

Gombossy, G. (2012, October 15). Does your complex allow political signs? Probably not. *CT Condo News* (Online). Retrieved from www.ctcondonews.com/does-your-complex-allow-political sign-lawn-signs

Gott, A. R. (2013). Ticky tacky little governments? A more faithful approach to community associations under the state action doctrine. *Florida State University Law Review, 40*, 201–220.

Grant, J., & Mittesteadt, L. (2004). Types of gated communities. *Environment and Planning B: Planning and Design, 31*, 913–930.

Harris, R. (2013). The American dream rebranded. *Reviews in American History, 4*, 717–722.

Kennedy, D. J. (1995). Residential associations as state actors: Regulating the impact of gated communities as nonmembers. *The Yale Law Journal, 105*, 761–793.

Lattman, P., & Haughney, C. (2011, February 1). Dakota co-op board is accused of bias. *The New York Times* (Online). Retrieved from www.nytimes.com/2011/02/02/ny region/02 dakota.html?pagewanted=all

Lloyd, C. (2009). The myth of "privatopia": Do private residential governments mean the end of the American dream? *SF Gate* (Online). Retrieved from http://sfgate.com/entertainment/article/the-myth-of-private-do residential-27-10581.php

Low, S. (2004). *Behind the gates: Life, security and the pursuit of happiness in fortress America.* New York, NY: Routledge.

Marsh v. Alabama, 326 U.S. 501 (1946).

McCabe, B. C. (2011, July/August). Homeowners associations as private governments: What we know, what we don't know, and why it matters. *Public Administration Review, 71*, 535–542.

McKenzie, E. (1994). *Privatopia: Homeowner associations and the rise of residential private government.* New Haven, CT: Yale University Press.

McKenzie, E. (2011). *Beyond privatopia: Rethinking residential private government.* Washington, DC: The Urban Institute Press.

Metromedia, Inc. v. San Diego, 454 U.S. 490 (1981).

O'Connor, D. (2012, October 3). Watchdog: Can a condo association ban political lawn signs. *Twin Cities.Com Pioneer Press* (Online). Retrieved from www.twincities.com/ci_21761427/watchdog-can-condo-association-ban-political-signs

Peter Lawfords barred from eastside co-op. (1964, April 7). *The New York Times* (Online). Retrieved from www.nytimes.com/1964/04/07/peter-lawfords-barred-from east-side-coop-html

Pieciak, M. S. (2009). My house, your rules: One court's timid retreat from its constitutional principles. *The Holy Cross Journal of Law and Public Policy, 13*, 147–174.

Princeton University v. Schmid, 455 U.S. 100 (1982).

Pruneyard Shopping Center v. Robbins, 447 U.S. 74, (1980).

Rahe, L. T. (2002). The right to exclude: Preserving the autonomy of homeowners' associations. *The Urban Lawyer, 34*(2), 521–552.

Reich, R. (1991, January 20). Secession of the successful. *New York Times Magazine*. Retrieved from http://www.nytimes.com/1991/01/20/magazine/secession-of-the-successful.html?pagewanted=all

Rosenberry, K. (1984). The application of federal and state constitutions to condominium cooperatives and planned developments. *Real Property, Probate and Trust Journal, 1*, 1–31.

Suarez, A. J. (2006). Covenants, conditions, and restrictions ... on free speech? First Amendment right in common-interest communities. *Real Property, Probate, and Trust Journal, 40*, 739–763.

Umberger, M. (2011, June 17). *The Chicago Tribune* (Online). Retrieved from http://www.chicagotribune.com

CHAPTER NINE

Contested Urban Space

Zoning Regulations as a Political Resource for Community Groups

KEVIN M. CARRAGEE
Suffolk University

This study focuses on the use of zoning regulations as a political resource by community groups in struggles over urban space. It connects with the expanding research literature concerning contested urban space. Conflicts over urban space represent enduring political struggles, struggles much influenced by political power (Barker, 2002; Carragee, 2005a, 2007; Zukin, 1996).

I examine a conflict over urban space in which a newly formed community group, the Presentation School Foundation (PSF), struggled with the Roman Catholic Archdiocese of Boston (RCAB) over the future of an elementary school that the Archdiocese planned to close; the school had served as a community anchor for eighty years in Boston's fragile Allston-Brighton neighborhood. Throughout the conflict, PSF relied exclusively on the commitment of its volunteers, given its lack of paid staff.

This three-year conflict involved the use of zoning regulations as a political resource; communication about zoning directly influenced the future purpose and communicative uses of the former Presentation school property. After initially not revealing its plans for the building, the RCAB announced that it would house administrative offices. The Boston Redevelopment Authority (BRA), the city's chief planning agency, held that the Archdiocese would need zoning relief for its plan given the building's changing use from educational purposes to administrative offices. This ruling reinforced the longstanding support of Mayor Thomas Menino for PSF's plan to purchase the building and transform it into a multi-service community center. The foundation seized the BRA ruling as a political opportunity in its campaign to oppose the RCAB and mobilized opposition to its request for zoning relief.

To further complicate the Archdiocese's plans, PSF petitioned the Boston Landmarks Commission (BLC) to designate the building as a historically significant structure. The Commission found that the building merited further study to determine its importance to Boston's history. This finding further complicated the Archdiocese's efforts to transform the building into administrative offices by mandating that the BLC review any proposal that would significantly modify the building.

The study, therefore, documents how zoning regulations and codes linked to historic preservation can be employed by community groups to oppose powerful institutions and to advance their own discourse concerning contested urban space. These tactics and others eventually forced the RCAB to sell the building to PSF in 2007. The PSF Community Center opened in 2012, eight years after the foundation's formation.

Figure 9.1. The community center. Photo by Liane Brandon.

This study advances a number of central arguments relating to urban communication. A significant form of urban communication involves contestation over urban spaces. Conflicts over urban space involve the symbolic production of meaning through communication and they can be studied through an analysis of the frames employed by social actors engaged in these struggles. Moreover, these conflicts are shaped considerably by zoning regulations because social actors engaged in these conflicts need to structure their public argumentation in light of existing zoning policies. The outcomes of these conflicts shape the communicative environment of cities. Power inequalities between groups engaged in urban communication also influence conflict over urban spaces. Despite this, this study demonstrates that underresourced groups, like PSF, can employ zoning regulations and codes relating to historic preservation to shape contested spaces in cities, influencing the communicative uses of these spaces.

Multiple perspectives, including research on contested urban space, engaged scholarship, and my experiences as PSF's board chair, shape this analysis.

The discussion begins with the research perspectives that inform this study. I first explore scholarship on contested urban space. I, then, examine how perspectives on engaged scholarship shaped my involvement with PSF. Following this discussion, I provide a brief introduction to the struggle between PSF and the Archdiocese over the building's future. This discussion sets the stage for an examination of how PSF activists successfully used zoning regulations and codes relating to historic preservation as political resources in their struggle over the building. The analysis of the controversy ends with a discussion of the conflict's key turning point: the Archdiocese strategic blunder to abruptly close Our Lady of the Presentation School, locking young children out of their school. I conclude with a call for communication scholars to focus more attention on discursive struggles over urban space, recognizing that these struggles are shaped by zoning and other land use regulations. I also encourage communication researchers committed to engaged scholarship to work with community groups in struggles over urban space.

Contested Urban Spaces and Discursive Struggles

Increasing scholarly attention has focused on contested spaces in cities and on the increasing privatization of public space (Barker, 2002; Davis, 1992; Zukin, 1996). The communication discipline, however, has made only a modest contribution to this research. Much of the research on contested urban spaces originates in other disciplines, including urban studies and sociology. While this scholarship has increased our awareness of places as arenas of political struggle, many studies have not provided detailed analyses of the urban communication produced by social actors engaged in contestation about urban space. By their very nature, discursive struggles over contested spaces in cities are a significant form of urban communication; these struggles frequently involve symbolic conflicts relating to zoning and other land use regulations.

Mitchell (2003) highlighted the recurring nature of conflicts concerning urban spaces. He emphasized that "different people with different projects must necessarily struggle with one another over the shape of the city" (p. 18). MacDonald (2011) stressed the ubiquity of "site fights" in the contemporary urban landscape, with contending groups struggling over the control and use of specific locations within cities. These site fights can involve multiple issues, including the gentrification of urban neighborhoods, the construction of affordable housing, and the expansion of non-profit institutions.

At its best, this research highlights the complex character of struggles over urban space. These struggles frequently correspond with broader conflicts in American society involving race, class, and gender; they often pit very resourced groups, including developers and political elites, against marginalized communities. Given the increasing influence of neoliberalism on urban policies and the inequitable distribution of power in American society, conflicts over urban space frequently end with victories by elite interests over marginalized communities (Davis, 1992; MacDonald, 2011; Mitchell, 2003). Lowes (2002) highlights the influence of "place entrepreneurs ... powerful coalitions of developers and political and business elites who are the principal architects of urban development" (p. 113).

Despite this general tendency, Gibson and Lowes (2007) remind us that grassroots opposition has, at times, "forced urban elites to make concessions that worked to the advantage of local communities" (p. 11). Berry (2003) documented the successful effort of a grassroots group in Convent, Louisiana, part of Cancer Alley, to block a powerful multinational corporation, initially supported by local and state government, from locating a large petrochemical plant in its community. He contended that this group succeeded because of its skill in articulating an environmental justice frame and because of its ability to forge an impressive coalition, including Greenpeace and the Sierra Club, opposed to the corporation's plans. Similarly, Martin (2004) chronicled the ability of a local community group in Athens, Georgia to secure a victory in a land use struggle involving a hospital's expansion plans. These and other cases studies reveal that grassroots groups, at times, can overcome the formidable resource inequalities they confront in struggles involving contested urban space.

Research has documented that social movements and reformers have used zoning regulations as a means to advance their goals. For example, in the early twentieth century, a coalition of reformers, including labor activists, public health advocates, social workers, progressive urban planners and housing advocates critical of tenement conditions, joined forces to produce the first zoning and housing codes in American cities, dramatically improving public health and housing conditions for the urban poor (Lopez, 2012). This coalition benefited from its relationship to muckraking journalists associated with the Progressive movement who exposed the shocking housing conditions confronting many city residents. Jacob Riis's compelling book, *How the Other Half Lives* (1971/1890), helped galvanize urban reform movements.

Other movements have used zoning codes to produce what they view as progressive urban reform, focusing their attention on environmental concerns, the revitalization of urban neighborhoods, the preservation and creation of affordable housing, and the creation of places that promote social interaction within cities.

The environmental movement, frequently working with public transit advocates, has influenced some cities to enact smart growth policies, including zoning regulations that encourage denser housing developments near public transportation. Environmental advocates also have shaped policies that preserve greenbelts around cities through urban growth boundaries (Addison, Shang, & Coomes, 2013).

Significantly influenced by Jacobs's views (1992/1961) on the characteristics of vital urban neighborhoods, reformers and urban planners have embraced mixed use zoning as a means to revitalize urban neighborhoods (Zukin, 1996). Mixed use zoning, integrating residential, commercial and leisure uses, enhances social interaction by the use of settings and streets by diverse people at diverse times.

In the face of rapidly escalating rental and housing prices in many American cities, housing advocates have tried to employ zoning and land use strategies as a means to preserve and expand affordable housing. These efforts frequently have not met their ambitious goals. Beitel (2007) documented the success of neighborhood activists in San Francisco to secure changes in their city's housing laws to protect the historical character of their neighborhoods from overdevelopment. Despite the accusations of the real estate industry that these zoning changes produced rising housing prices, Beitel's longitudinal study found no relationship between these changes and increased housing costs. Nonetheless, he concluded that only massive public subsidies for the creation of affordable housing can ease the city's housing crisis.

MacDonald (2011) examined a complex effort by a community and labor union coalition to employ land-use strategies and zoning to modify the planned redevelopment of Brooklyn's Coney Island. An alliance between a hotel workers union and local community groups sought to win concessions related to affordable housing and worker rights as part of a city-sponsored plan to redevelop Coney Island to accommodate hotels and market-rate housing. While the coalition obtained a guarantee that the planned hotels would employ unionized workers, it failed to secure its demands for affordable housing. This outcome reflected the union's position as the dominant coalition partner. The union and community groups operated in a neoliberal urban policy environment in which the Bloomberg administration, aligned with developers, gentrified working class neighborhoods.

Confronted with the lack of funds to create new urban parks, some cities have created zoning incentives allowing for greater building density in return for the creation and maintenance of privately owned public spaces, including plazas and arcades. New York City has led in this effort to create spaces that enhance social interaction and create a more inviting streetscape. The results of this effort, however, are mixed. While these incentives have created more than 500 privately owned public spaces and produced some spaces of impressive quality, many of these spaces are poorly designed (Kayden, 2000). These failed spaces illustrate Whyte's observation (1988) that "it is difficult to design a space that will not attract people. What is remarkable is how often it has been done" (p. 109).

The communication discipline has made only a modest contribution to research on contested urban space. Sadly, communication researchers have neglected urban issues, failing to recognize that these issues are tied to broader political and economic contexts (Gibson & Lowes, 2007). In recent years, a growing body of communication research on urban issues, including contested urban space, has been produced, in part, through the efforts of the Urban Communication Foundation. Drucker and Gumpert (1991, 1996) have highlighted how zoning regulations and other codes shape the communicative environment of cities.

Communication as a discipline has much to contribute to scholarship on contested urban space despite its limited attention to this issue thus far. While spaces have a material reality, their social meanings are constructed through communication. Struggles over spaces and urban design are discursive in nature and communication as a discipline historically has focused on signification as a social process and on symbolic struggles. Scholarly examinations of these struggles reveal their connection to significant issues in American society, including the distribution of power and continuing inequalities. Barker (2002) reminds us that "cities are terrains of contestation in which various ethnic, class, gender and organizational agents struggle to shape the social and built environment" (p. 370).

In a perceptive recommendation focusing on communication research related to contested urban space in particular and urban issues in general, Gibson and Lowes (2007) highlight the need for scholarship that "takes issues of meaning and culture seriously, while embedding this analysis in its material and economic context" (p. 17); they endorse research that integrates political and economic analysis with a careful examination of urban communication. The economic, political and symbolic resources available to social actors influence struggles over urban space, including conflicts related to zoning. Therefore, corporate and pro-development elites, frequently supported by city governments, often obtain their goals, but marginalized communities obtain victories in these conflicts when they are able to effectively mobilize opposition to particular uses of urban space and simultaneously advance an alternative discourse that attracts support (Carragee, 2005a, 2007; Gibson & Lowes, 2007; Lowes, 2002).

A number of exemplary case studies in the communication literature on contested urban space exist, illustrating the benefits of integrating political and economic analysis with careful attention to the discourses that define conflicting visions of urban space. For example, Buttny (2010) and Buttny and Cohen (2007) provide discourse analyses of public hearings in which community groups voiced their opposition to a zoning change sought by Wal-Mart to construct a super store. Lowes (2002) documented the opposition of a coalition of community groups in Vancouver, Canada that prevented a major auto race from using a park as a venue. This grassroots victory occurred despite the significant disparity between the resources available to the corporate and governmental elites seeking to locate the auto race within the park and the community groups. Lowes located his study within a broader research literature that has focused on cities using spectacular entertainment or sporting events as "signifiers of civic prosperity and ambition" (p. 116).

In contrast, Broad (2013) examined the ability of corporate and political power to triumph in a protracted land-use struggle over a contested urban space, the South Central Farm in Los Angeles, that for a time was the largest urban garden in the United States. By applying Carey's distinction between the transmission and ritual views of communication to this struggle, he documents how the South Central Farmers employed a discourse associated with the ritual view, highlighting the garden's role in sustaining community. In contrast, government officials and corporate interests employed a discourse consistent with the transmission view, highlighting market-based arguments that the land could be put to more productive economic use.

While scholars can investigate discursive struggles over contested urban spaces from multiple perspectives, a promising approach links these struggles to framing and how this process is central to meaning production.

Frames organize discourses by their patterns of selection, emphasis, and exclusion. By framing issues, including struggles related to contested urban space, social actors define what is and what is not relevant to an issue. Entman (1993, p. 52) writes that frames "select some aspects of a perceived reality and make them more salient in a communicating text." Considerable research has examined frames in news discourses (Carragee & Roefs, 2004; Gamson, 1992; Ryan, 1991).

The discursive struggles over spaces in cities involve framing contests in which competing social actors advance their definitions of the current character and the future of these spaces. These struggles take place in specific zoning and land use regulatory environments depending on the city in which they occur, directly shaping the urban communication produced by sponsors of particular frames. These frame sponsors seek to dominate discourses about these places, including public policy discourses and news discourses. The domination of the symbolic arena often leads to the domination of material space. Given the importance of cultural and material resources (money and staff, for example) in framing processes, elites tend to dominate discourses about contemporary issues, including debates about contested urban space (Carragee, 2005a, 2007; Carragee & Roefs, 2004; Gamson, 1992).

Nonetheless, attention to framing helps us to understand how, under certain circumstances, community groups and social movements succeed in mobilizing support and in shaping public policy or outcomes relating to controversies, including disputes involving contested urban space and zoning regulations. Contestation about urban spaces, particularly conflicts involving marginalized groups struggling against elite definitions of urban space, involves collective action (Carragee, 2005a, 2007).

Engaged Research and Contested Urban Space

While past research on contested urban space informs this analysis, engaged scholarship also shapes this study. The perspectives advanced by Communication Activism Research (CAR) and the Media Research and Action Project (MRAP) influenced my involvement with PSF; CAR and MRAP stress the significance of collective action in securing political change.

CAR involves communication scholars using their theories and knowledge to work with and for underresourced groups to intervene into unjust discourses and material conditions to make them more just. CAR involves scholars engaging in activism, but also analyzing their involvement with marginalized groups seeking political change (Frey & Carragee, 2007a, 2007b, 2012).

MRAP, a group of Boston-based social scientists and activists that I joined in 1990, assists community groups in framing political issues, challenging obstacles to news media access, and developing integrated organizing and media strategies to obtain progressive change. Since its founding in 1986, MRAP has employed social movement theory and framing theory to guide more than 300 interventions with community groups (Ryan, Carragee, & Meinhofer, 2001; Ryan, Carragee, & Schwerner, 1998). MRAP's emphasis on framing played a decisive role in structuring my involvement with PSF and in shaping this analysis.

The influence of engaged research on this study takes on added significance given the lack of engaged scholarship that particularly focuses on urban communication and contested spaces in cities. Abraham and Maney (2012) helped two marginalized groups challenge NIMBY opposition to the siting of a domestic violence shelter in New York City and a Latino immigrant-hiring center on Long Island. Carragee (2005a) analyzed his intervention with an Allston-Brighton community group that helped relieve Boston's housing crisis by influencing Boston College to house more of its undergraduate students on campus.

The Context of the Presentation Struggle

The RCAB's clergy sexual abuse scandal produced the controversy over the Presentation School building as contested urban space. Reeling from the scandal's scope, the Archdiocese announced a major consolidation plan in May 2004, closing 64 parishes and a number of Roman Catholic schools, including Our Lady of the Presentation. Extensive news reporting, including award-winning coverage by *The Boston Globe*, revealed that Archdiocesan officials, including Cardinal Law, knowingly transferred priests engaged in the sexual abuse of minors. Archbishop O'Malley, who succeeded Law, agreed to an $85 million settlement with more than 500 victims of clergy sexual abuse.

The sexual abuse scandal represented a profound shock for Boston-area Roman Catholics, sparking collective action. A Roman Catholic lay group, Voice of the Faithful, emerged, demanding structural reforms within the Archdiocese. Opposition to specific church closings developed, with protesters in 2004–2005 seizing eight churches in round-the-clock occupations.

In May 2004, parents of students at the Presentation School and community residents organized protests against the RCAB's decision to close the school, demanding a one-year extension of the school's life. These protests attracted news coverage and political support. Because of this pressure, the Archdiocese granted the school community a one-year extension. Significantly, key activists involved in securing this extension viewed this demand as a tactical move that would allow time to develop a plan for the building's future.

The extension provided a window of opportunity for residents who sought to extend the school's life. In June 2004, community activists formed PSF, seeking to maintain the building as a community anchor. Between June and December 2004, the foundation developed a comprehensive plan for the building's future. PSF sought to purchase the building at full-market value and transform it into a multi-service community center, serving children and families. Mayor Thomas M. Menino and the entire Boston City Council endorsed the plan along with twenty-five civic groups and non-profit organizations in Allston-Brighton, including the YMCA, WGBH, and Boston University.

Despite the breadth of this support, the Archdiocese rejected the 150-page plan in a curt five-sentence letter in January 2005. The Archdiocese said that it would retain ownership of the building. The rejection of PSF's plan triggered a prolonged conflict between the foundation and the RCAB.

Using Zoning Regulations and Codes Relating to Historical Preservation as Political Resources

PSF activists viewed zoning regulations and codes concerning historic preservation as political resources that could be employed in the struggle to retain the building as a community anchor. These regulations and codes, in their view, could be used to complicate the Archdiocese's plans for the building's future; they also could be used to convince potential developers that a purchase of the building and its transformation to expensive housing would face widespread community opposition and daunting challenges related to zoning and historic preservation.

PSF employed zoning regulations and codes relating to historic preservation as political resources within the context of a much broader strategy that shaped its conflict with the RCAB. This strategy employed multiple tactics, including demonstrations, street theatre, Boston City Council hearings, and efforts to attract and influence news media coverage of the controversy (Carragee, 2011). These diverse tactics placed considerable pressure on the RCAB.

The forging of a broad coalition supporting PSF and the expertise of its board members on multiple issues dramatically increased the social and political capital available to the foundation in its struggle against the Archdiocese. The foundation's board included parents of children attending the school, elected officials, representatives of major non-profit organizations in the community (the YMCA and WGBH, for example) and business owners. The formation of the foundation and the creation of a broad coalition supporting PSF represented a decisive organizing accomplishment because it profoundly expanded the resources and expertise available to the effort to maintain the building as a community anchor. The passion of parent activists was linked to the knowledge and social networks of seasoned community activists, long-term executive directors of non-profit institutions, and elected officials.

This expanding social and political capital provided the foundation with considerable expertise relating to zoning and historic preservation. The board included four elected officials, a Massachusetts State Senator, two Massachusetts State Representatives and a Boston City Councilor, all of whom had considerable knowledge of Boston's zoning regulations. The Secretary of State of Massachusetts, William Galvin, a graduate of the Presentation School, helped create the foundation and supported its efforts, increasing PSF's political capital and its expertise relating to zoning and historic preservation.

While these board members brought needed expertise and political capital to PSF, the board also included a long-term community activist, Charlie Vasiliades, who had extensive experience in *grassroots campaigns* involving zoning and historic preservation. A member of many community organizations, Vasiliades had shaped Allston-Brighton zoning regulations, and engaged in multiple grassroots struggles involving zoning, including a successful campaign, lasting twenty years, to prevent an overly dense housing development. He also had played a key role in efforts to designate Allston-Brighton buildings as Boston landmarks.

The Archdiocese's rejection of PSF's proposal to purchase the building sparked three demonstrations in late January and early February; these demonstrations attracted considerable local news attention and a lengthy CNN story. In this early coverage, PSF activists criticized the RCAB for failing to indicate its planned use of the building after the school's closing. In the face of this criticism, the Archdiocese announced that it would re-locate its ecclesiastical court to the building. PSF supporters viewed this announcement with deep skepticism, believing that it was a cover for the RCAB's plan to sell the building for maximum profit to a developer who would transform it into luxury condominiums.

The RCAB's public commitment to re-locating its ecclesiastical court to the building produced increased attention by PSF to the use of zoning regulations as a political resource. Simply put, if the City of Boston determined that the use of the school as an office building represented a *change in use*, then, the RCAB would need a zoning variance to house its court in the building. This would provide PSF and its political allies an opportunity to mobilize support against the granting of a zoning variance by the Zoning Board of Appeals.

The potential use of zoning regulations to influence the RCAB's decisions concerning the disposition of closed Church properties in Boston had emerged as a public issue prior to the RCAB's announcement of its intended use of the building. In a Boston City Council hearing on the "Future Zoning and Land Uses of Buildings and Properties Owned by the Roman Catholic Archdiocese of Boston," held in October 2004, City Councilors criticized the Archdiocese for its failure to enter into a master planning process with the BRA concerning the future uses of closed churches and schools. City Councilor Jerry McDermott, representing Allston-Brighton and a PSF Board member, noted that anticipated closings of churches and schools raised significant zoning issues. Secretary of State Galvin testified that the "use of these properties is regulated by the zoning code," stressing that potential buyers should take note of zoning regulations linked to Church properties.

Another Boston City Council hearing, held on February 17, 2005, represents a key moment in linking the controversy over the building to zoning regulations. This hearing deserves a careful analysis because of multiple factors, including the framing of the controversy by PSF representatives providing testimony and the

BRA finding that the Archdiocese would need a zoning variance to re-locate its ecclesiastical court to the building.

PSF Board Members Jennifer Doyle, Tom O'Brien and I represented the foundation at the hearing. Our combined testimony expressed the three major frames that characterized PSF's urban communication during its conflict with the RCAB: a best use frame; a threatened community frame; a broken vow frame. Because of their significant role in advancing PSF's definition of the building as contested urban space, I briefly define them.

The best use frame contrasted the proposed uses of the building by the Archdiocese and PSF. It stressed the foundation's effort to establish a multi-service community center, positively defining this use by contrasting it with the RCAB's effort to re-locate its ecclesiastical court, an agency focusing primarily on granting marriage annulments, to the building. Doyle and O'Brien advanced this frame during the hearing.

Significantly, PSF's urban communication relating to the best use frame corresponded to the zoning controversy involving the property. The foundation's proposal for a community center represented a use aligned with the property's existing zoning for educational purposes, while the RCAB's proposed use violated the existing zoning, thereby, demanding a variance. Zoning regulations, therefore, directly shaped PSF's urban communication.

The threatened community frame defined Allston-Brighton as a fragile neighborhood, experiencing a loss of working and middle class families because of rapidly escalating home prices and rents and experiencing unprecedented expansion by Harvard University and Boston College (Carragee, 2005a). This frame also highlighted that 23 percent of Allston-Brighton residents lived in poverty. Doyle articulated this frame in depth in her testimony, stressing that Allston-Brighton was a "community at risk."

The broken vow claim contrasted the Archdiocese's rejection of PSF's offer to purchase the building with its repeated claim that it would evaluate proposals for the future use of Church-owned properties by taking into account its economic needs and the degree to which proposals corresponded with the Roman Catholic Church's social mission. In its proposal to the Archdiocese and in its subsequent public statements, PSF highlighted that it had advanced a market-based offer for the building and that its planned uses of the building, focusing on services for children and families, were entirely consistent with the Church's mission. O'Brien advanced this frame, accusing the Archdiocese of violating its own "principles" in rejecting the foundation's proposal.

The council hearing repeatedly focused on zoning as a means to protect Boston neighborhoods potentially harmed by the closing of Roman Catholic churches and schools. Councilor Hennigan stated that zoning regulations could be used as "leverage" against the Archdiocese, while Councilor Ross concluded, "the only thing that protects us is zoning."

In a response to a question involving zoning issues relating to the contested Presentation School property, Tom Miller, the BRA's Director of Economic Development, stated that the re-location of the Archdiocesan court to the building would "require zoning relief" given the BRA finding that the city had zoned the property for educational uses, and that the court represented an office use. This was the hearing's decisive moment from PSF's perspective because it provided a major political resource to employ against the RCAB.

The BRA's conclusion that the Presentation school property was zoned for educational purposes had important implications for the future communicative uses of this urban space. This decision supported the uses proposed by PSF, preserving the space for community and educational uses and, therefore, directly shaping the social interaction and communication that would take place in the building. This action by the BRA made it far less likely that the space could be used for either Archdiocesan administrative purposes or for the development of luxury housing, alternatives that would, in turn, produce a space for communication focused primarily on private concerns.

The BRA's finding that the RCAB would need a zoning variance takes on added significance given Boston's political culture, especially the degree of power historically exercised by its mayors. In his long tenure as Boston's mayor, eventually spanning 21 years, Mayor Menino exercised considerable control over the BRA and the Zoning Board of Appeals. The RCAB, therefore, faced the likelihood that Mayor Menino would join other elected officials along with PSF activists to oppose any Archdiocesan effort to obtain a zoning variance.

O'Brien was quick to exploit this opportunity, concluding that it would be in the best interests of the Archdiocese to partner with PSF, rather than risk a protracted zoning struggle. He also broadened his argument by connecting the struggle over the Presentation School building to zoning conflicts involving other church-owned properties: "They [the RCAB] will find battlezones throughout the city on every property." O'Brien claimed that these likely zoning challenges would reduce the value of Archdiocesan properties given the reluctance of buyers to purchase properties in the face of both potentially strong opposition and the need to obtain zoning variances to allow changes in the use of these properties.

PSF exploited the BRA finding by communicating directly with the RCAB on its willingness to use zoning regulations as a resource in its conflict with the Archdiocese and by attracting news coverage to this issue. In a March 14, 2005 letter to Archbishop O'Malley on PSF's behalf, I wrote that the foundation had:

> sought a cooperative resolution to this controversy, but now we stand ready for a protracted conflict. We remain fully committed to our proposal ... we will oppose any

> effort by the Archdiocese to obtain the zoning relief necessary to relocate its offices to the Presentation School building and will also oppose any change in zoning for the 4 acres of Presentation parish.

PSF included the extension of the zoning conflict to the parish property in an effort to bring increased pressure on the Archdiocese to reach an agreement with the foundation concerning the school property. PSF activists and local elected officials expected that the parish property would be sold to a developer committed to constructing housing. A potential lengthy and combative zoning struggle likely would decrease interest in the parish property by developers and, therefore, reduce the price the RCAB could obtain for the property. This tactical move represented an effort to leverage zoning restrictions as a means to influence the Archdiocese to reach an agreement with PSF.

Following the Boston City Council hearing and the letter to Archbishop O'Malley, PSF successfully influenced news coverage to highlight its intention to oppose the zoning relief necessary for the Archdiocese to re-locate its ecclesiastical court to the building. From its inception, PSF recognized the news media's importance as a public forum to draw attention to its struggle and to place additional pressure on the RCAB. The zoning conflict provided an opportunity to attract news coverage and to advance PSF's best use frame.

The release of the foundation's letter to Archbishop O'Malley resulted in news stories in *The Boston Globe*, *The Allston-Brighton Tab* and the *Allston-Brighton Bulletin* (Paulson, 2005; Uhlin, 2005). The *Bulletin* story emphasized PSF's commitment to a lengthy struggle for the contested building.

In an effort to attract news coverage prior to a planned demonstration timed to the school's closing in June 2005, PSF attracted additional news coverage that noted the zoning controversy. This coverage included stories in the *Boston Herald* and *The Allston-Brighton Bulletin,* and op-ed columns that I wrote for the *Allston-Brighton Tab* and the *Allston-Brighton Bulletin* (Atkins, 2005; Carragee, 2005b, 2005c; Gelzinis, 2005; Kaminski, 2005).

During this period, PSF's most successful effort to influence news coverage resulted in a *Boston Globe* opinion column that appeared a week before the school's closing. While employing the best use and broken vow frames, Michael Jonas (2005) emphasized the zoning controversy:

> … a change in use will require zoning relief from the city, setting the stage for the battle to move to City Hall, where the Archdiocese may find few friends.
>
> "I think there could have been a win-win, and it didn't happen," says state Representative Kevin Honan. "I would expect serious trouble at the zoning board." Carragee says of any zoning relief the church may seek, "We'll oppose that with all our strength." (p. 8)

News coverage of the zoning controversy represented another form of pressure on the RCAB.

Concurrent with initiatives related to zoning, PSF activists began a process that would designate the Presentation School building as a Boston landmark, providing another example of how broader regulatory frameworks in cities shape urban communication over contested space. PSF embarked on this effort to frustrate Archdiocesan efforts to renovate the building or any potential sale of the building to a developer interested in transforming it into housing. The landmark petition, developed by Vasiliades, argued that the building was a "distinct Brighton landmark" and stressed its important educational role in Brighton's history. It also emphasized the building's architectural importance, noting that a prominent Boston architectural firm designed the building and pointing out that the building is one of the "few remaining examples of mission style architecture in the area." It highlighted that the building met the criteria for inclusion in the National Register of Historic Places.

The BLC in April 2005 "voted to accept the petition to designate Our Lady of the Presentation School … as a Boston Landmark." This represented a significant first-step in the long process of designating the building as a historic landmark. The Commission added the building to a lengthy list of properties requiring the preparation of a study that would assess whether it met the criteria for landmark designation.

Because of Vasiliades's expertise relating to the process of securing landmark status, PSF recognized that Boston's landmark designation process extends over a lengthy period, owing to the BLC's small staff and limited budget. These limitations restrict the Commission to conducting only three reports each year that evaluate specific properties in terms of the criteria used for landmark designation; therefore, a lengthy list of properties builds up over time awaiting formal review. At the time of this writing in 2016, the BLC has not yet conducted its formal study of the building as a potential Boston landmark.

Despite these complexities, the BLC's decision to accept the petition to designate the building as a Boston landmark still afforded PSF specific advantages. It provided a degree of protection to the building during the lengthy delay before a formal study, making it difficult to demolish the building or to extensively change its appearance. Also, a pending landmarks review would make the property less attractive for potential buyers eager to transform the building into housing. Finally, the BLC has the ability to modify the sequencing of the review process when it perceives that a building faces an immediate threat given the sale of the property or given a proposed renovation that endangers the building's historic character. These protections, albeit limited in scope and dependent on the BLC's actions, still possessed decided advantages. They complicated any potential RCAB effort to renovate the building to house administrative offices or any potential sale of the property to a developer.

Transforming the Struggle into a Crisis

The Archdiocese transformed its struggle with PSF into a crisis by its ill-advised decision to abruptly close the Presentation School two days before its scheduled permanent closing, locking children out of their school on June 8, 2005. In explaining its action, the RCAB cited fears of an occupation of the school by residents. This decision cancelled graduation ceremonies for students. Angered by the lockout, PSF organized a graduation ceremony for kindergartners in a public park across from the school and began camping in that park to signal its determination to continue its struggle to purchase the building.

The lockout proved to be a significant strategic blunder for the Archdiocese. It provoked intense community opposition and strident criticism from Boston politicians. Mayor Menino organized a graduation ceremony at historic Faneuil Hall. The lockout and subsequent protests attracted extensive local news coverage, including four page one stories in *The Boston Globe* over a six-day period as well as national and international reportage by, for example, *The New York Times*, *The Los Angeles Times*, NPR, and *The Economist.*

Five days after the sudden school closing, the Archdiocese reversed its long-standing position and negotiated an agreement with PSF representatives. The agreement emphasized that both parties would make a good faith effort to reach a settlement leading to the building's sale to the foundation.

After eighteen months of difficult negotiations, the two parties reached an agreement, with the Archdiocese reducing its sale price by $1 million (PSF had offered $2 million) as an effort to repair its relationship with the Allston-Brighton community. The foundation gained building ownership in June 2007, and, after a $5.6 million renovation, opened the community center in May 2012. The center offers multiple services, including an affordable pre-school, an after-school program, health services, anti-poverty programs, and transportation services. In the midst of a severe economic recession, PSF volunteers accomplished these milestones by raising $4.2 million, including $1.25 million in historic tax credits from the Commonwealth of Massachusetts to preserve the character of the building's exterior. Eight years after its founding, PSF had regained the building as a community anchor, transforming it from contested urban space into a community resource.

Conclusion

While the lockout and its consequences played an essential role in helping PSF secure building ownership, the foundation's repertoire of tactics created a context in which the Archdiocese made an important strategic error. The use of zoning regulations and codes linked to historic preservation as political resources contributed to PSF's overall strategy to use multiple means to regain the building as a community anchor.

The use of these regulations and codes opened the possibility for a long standoff concerning the building's future. PSF activists accepted the possibility that the building would be empty for a lengthy time as struggles relating to zoning and historic preservation continued. They much preferred this outcome to the re-location of Archdiocesan offices to the building or, in their view the more likely result, the building's sale to a developer eager to convert the building into expensive housing. Even after the lockout and its likely favorable consequences for the foundation's ability to purchase the building, the potential use of zoning regulations and codes involving historic preservation remained as leverage against the Archdiocese during PSF's difficult negotiations with the RCAB.

This study highlights that a significant form of urban communication focuses on contested urban space. It also underscores the benefits of examining struggles over urban space by carefully evaluating the construction of meaning by social actors engaged in these struggles. These actors advance competing definitions of urban spaces and their potential uses. Given its traditional focus on signification, the communication disciple has much to contribute to a more detailed understanding of conflicts over urban space, including zoning conflicts. This study demonstrates how communication focusing on zoning issues shaped the future communicative uses and social interaction within the former Presentation School building.

This focus on the production of meaning needs to be integrated with an examination of the political and economic context in which these struggles take place. This analysis needs to include an evaluation of the resources available to social actors engaged in contestation about urban spaces. Research on contested urban space needs to consider how power shapes these struggles.

Zoning, land use regulations and codes relating to historic preservation directly shape urban communication focusing on contested space in cities. Discourses about contested urban space are shaped by these regulations and codes, with advocates presenting conflicting definitions of the future use of urban spaces advancing diverging interpretations of these regulations and codes or seeking changes in these regulations and codes.

This study also indicates the utility of applying a framing approach to struggles over urban space. Through its careful and consistent framing of the controversy, PSF mobilized community and political opposition against the Archdiocese. Boston's zoning regulations and its codes relating to historical preservation directly influenced PSF's urban communication and its framing of the controversy. Framing research can enrich our understanding of struggles over urban space given its attention to the construction of meaning and to the resources available to frame sponsors.

In this struggle, PSF benefited from both the zoning of the property for educational use and from Boston's regulatory framework involving historical preservation. This case study, therefore, reveals the significance of zoning practices that reserve particular urban spaces for community or educational use, protecting them from competing uses, many of which return considerable profit to developers. Significantly, PSF had the cultural capital and expertise necessary to employ zoning regulations and codes relating to historical preservation as political resources.

Communication scholars confront an urban landscape rich in possibilities for future research on contested urban space. Possibilities include studies focusing on conflicts relating to the gentrification of urban neighborhoods, the privatization of public space, the siting of social services (for example, homeless shelters and drug treatment facilities), environmental justice issues confronting poor and minority neighborhoods, and the expansion of non-profit institutions into neighborhoods. Zoning, land use regulations and other urban regulatory policies, including codes relating to historical preservation, inevitably influence these controversies.

In particular, I highlight the need for communication scholars to grapple with conflicts relating to the development of affordable housing in American cities, especially given increasing income inequality and its implication for cities. Many cities, including New York, San Francisco and Boston, are confronted with severe housing crises. Escalating rents and home prices have forced working and middle class residents from these and other cities. The development of affordable housing for working class and middle class residents of cities remains an urgent need, but opposition to this housing is widespread, producing prolonged conflicts over the need for and location of affordable housing.

For example, Mayor Bill de Blasio's 10-year plan to build or preserve 200,000 affordable apartments in New York City faces opposition focusing on diverse concerns, including the need to rezone neighborhoods to allow for greater housing density and fear of increased gentrification of poor and minority neighborhoods given new housing development. For his part, Mayor de Blasio links his plan to safeguarding New York as a mixed-income city (see the *New York Times*'s editorial "Saving a Mixed-Income New York," 2016).

This struggle in New York City and similar conflicts over the need for and location of affordable housing provide a dramatic opportunity for studies that focus on urban communication and competing discourses on affordable housing and on specific "site fights" related to the location of this housing. In addition, communication scholars should help inform the development of zoning policies and land use regulations that assist in the development of affordable housing in an effort to ensure that American cities and particular neighborhoods safeguard their economic diversity.

Controversies linked to affordable housing and to other struggles related to contested urban space also provide rich opportunities for communication scholars committed to engaged research. Conflicts over urban space frequently involve struggles between politically and economically advantaged and disadvantaged groups. Moving beyond scholarly analyses of urban communication focusing on contested urban space, engaged communication researchers can assist marginalized communities in their efforts to secure social justice within an urban environment too often dominated by elite interests. These activist researchers can help marginalized communities challenge zoning practices and land use regulations that serve elite interests. The perspectives advanced by CAR and MRAP can inform this research.

This study illustrates the advantages of conducting engaged communication research focused on contested urban space. My knowledge related to framing issues, mobilizing collective action, and influencing news media coverage benefited PSF. My contributions, however, were embedded in the collective work of many passionate volunteers who brought diverse skills to a good fight and overcame formidable challenges. They transformed a contested urban space into a site where community is produced and sustained. This study, therefore, is dedicated to them.

References

Abraham, M., & Maney, G. (2012). Transforming place and belonging through action research, community practice, and public policy: Comparing responses to NIMBYism. *Current Sociology, 60*(2), 178–201. doi: 10.1177/00113429220

Addison, C., Shang, Z., & Coomes, B. (2013). Smart growth and housing affordability: A review of regulatory mechanisms and planning practices. *Journal of Planning Literature, 28*(3), 215–257. doi: 1177/0885412212471563

Atkins, K. (2005, June 3). Group up in arms over school's use for perv inquisition. *The Boston Herald*, p. 8.

Barker, C. (2002). *Cultural studies: Theory and practice*. London: Sage.

Beitel, K. (2007). Did overzealous activists destroy housing affordability in San Francisco? A time-series test of the effects of re-zoning on construction and home prices, 1967–1998. *Urban Affairs Review, 42*(4), 741–756. doi: 10.1177/1078087406296795

Berry, G. (2003). Organizing against multinational corporate power in Cancer Alley: The activist community as primary stakeholder. *Organization & Environment, 16*, 3–33. doi: 10.1177/1086026602250213

Broad, G. (2013). Ritual communication and use value: The South Central Farm and the political economy of space. *Communication, Culture & Critique, 6*, 20–40. doi: 10.1111/cccr.12003

Buttny, R. (2010). Citizen participation, metadiscourse, and accountability: A public hearing on a zoning change for Wal-Mart. *Journal of Communication, 60*, 635–659. doi: 10.1111/j.1460-2466.2010.01507.x

Buttny, R., & Cohen, J. (2007). Drawing on the words of others: Zoning, Wal-Mart, and the threat to the aquifer. *Language and Society, 36*, 735–756. doi: 10.1017/S0047404507070674

Carragee, K. M. (2005a). Housing crisis: Gaining standing in a community coalition. In D. Croteau, W. Hoynes, & C. Ryan (Eds.), *Rhyming hope and history: Activists, academics, and social movement scholarship* (pp. 79–96). Minneapolis, MN: University of Minnesota Press.

Carragee, K. M. (2005b, May 20). OLP: Why the struggle continues. *The Allston-Brighton Tab,* p. 10.

Carragee, K. M. (2005c, June 2). To remember is to resist. *The Allston-Brighton Bulletin,* p. 5.

Carragee, K. M. (2007). Public space and the public sphere: Boston's City Hall Plaza as contested public space. In G. Burd, S. Drucker, & G. Gumpert (Eds.), *The urban communication reader* (pp. 105–128). Cresskill, NJ: Hampton Press.

Carragee, K. M. (2011). *Community mobilization and resistance in a fragile Boston neighborhood.* Paper presented to the International Communication Association Convention, Boston, MA.

Carragee, K. M., & Roefs, W. (2004). The neglect of power in recent framing research. *Journal of Communication, 54*(2), 214–253. doi: 10.1111/j.1460-2466.2004.tb02625.x

Davis, M. (1992). *City of quartz: Excavating the future in Los Angeles.* New York, NY: Vintage.

Drucker, S., & Gumpert, G. (1991). Public space and communication: The zoning of public interaction. *Communication Theory, 1*, 294–310.

Drucker, S., & Gumpert, G. (1996). The regulation of public social life: Communication law revisited. *Communication Quarterly, 44*(3), 280–296.

Entman, R. (1993). Framing: Toward a clarification of a fractured paradigm. *Journal of Communication, 43*(4), 51–58. doi: 10. 1111/j.1460-2466.1993.tb01304x

Frey, L. R., & Carragee, K. M. (Eds.). (2007a). *Communication activism: Vol. 1. Communication for social change.* Cresskill, NJ: Hampton Press.

Frey, L. R., & Carragee, K. M. (Eds.). (2007b). *Communication activism: Vol. 2. Media and performance activism.* Cresskill, NJ: Hampton Press.

Frey, L. R., & Carragee, K. M. (Eds.). (2012). *Communication activism: Vol. 3. Struggling for social justice amidst difference.* New York, NY: Hampton Press.

Gamson, W. (1992). *Talking politics.* New York, NY: Cambridge University Press.

Gelzinis, P. (2005, June 8). Church's social mission is lost in reconfiguration. *The Boston Herald,* p. 19.

Gibson, T., & Lowes, M. (2007). Introduction: The city as production, text, context. In T. Gibson & M. Lowes (Eds.), *Urban communication: Production, text, context* (pp. 1–21). Lanham, MD: Rowman & Littlefield.

Jacobs, J. (1992/1961). *The death and life of great American cities.* New York, NY: Vintage.

Jonas, M. (2005, June 5). Church's poor presentation: Archdiocese bucks neighborhood hopes. *The Boston Globe/City Weekly,* p. 8.

Kaminski, R. (2005, June 2). OLP rezoning not an option. *The Allston-Brighton Bulletin,* pp. 1, 8.

Kayden, J. (2000). *Privately owned public space: The New York City experience*. New York, NY: Wiley.

Lopez, R. (2012). *Building American public health: Urban planning, architecture and the quest for better health in the United States*. New York, NY: Palgrave Macmillan.

Lowes, M. (2002). *Indy dreams and urban nightmares: Speed merchants, spectacle, and the struggle over public space in the world-class city*. Toronto: University of Toronto Press.

MacDonald, I. (2011). Bargaining for rights in luxury city: The strategic dilemmas of organized labor's urban turn. *Labor Studies Journal, 36*(2), 197–220. doi: 10.1177/0160449X11404075

Martin, D. (2004). Reconstructing urban politics: Neighborhood activism in land-use change. *Urban Affairs Review, 39*(5), 589–612. doi: 10.1177/1078087404263805

Mitchell, D. (2003). *The right to the city: Social justice and the fight for public space*. New York, NY: Guilford Press.

Paulson, M. (2005, May 8). Resistance widens to parish closings: After uneasy calm, protests resume. *The Boston Globe*, p. A. 1.

Riis, J. (1971/1890). *How the other half lives: Studies among the tenements of New York*. New York, NY: Dover.

Ryan, C. (1991). *Prime-time activism: Media strategies for grassroots organizing*. Boston, MA: South End Press.

Ryan, C., Carragee, K. M., & Meinhofer, W. (2001). Theory into practice: Framing, the news media and collective action. *Journal of Broadcasting & Electronic Media, 45*, 175–182. doi: 10.1207/s15506878jobem4501_11

Ryan, C., Carragee, K. M., & Schwerner, C. (1998). Media, movements, and the quest for social justice. *Journal of Applied Communication Research, 26*, 165–181. doi: 10.1080/00909889809365500

Saving a Mixed-Income New York. (2016, March 12). *The New York Times*, p. A18.

Uhlin, E. (2005, March 24). Zoning could be next OLP battleground. *The Allston-Brighton Bulletin*, pp. 1–2.

Whyte, W. (1988). *City: Rediscovering the center*. New York, NY: Doubleday.

Zukin, S. (1996). *The culture of cities*. Malden, MA: Blackwell.

PART THREE

Manner and Regulation

CHAPTER TEN

The Urban Broadband Revolution

What Cities Can Do to Bolster Connectivity

CHARLES M. DAVIDSON
ACLP at New York Law School

MICHAEL J. SANTORELLI
ACLP at New York Law School

Cities are critical hubs of technological innovation in the 21st century. This is not surprising given the prominent role that urban environments have long played in driving economic creation and supporting technological progress (Glaeser, 2011; Townsend, 2013). The characteristics of cities that have facilitated these gains historically—the clustering of well-educated workers; demographic diversity; ready access to important economic inputs like cutting-edge communications tools; and relative prosperity, among many others—have become even more pronounced in recent years, stoking optimism about the singular role that urban areas are poised to play in revolutionizing how key public policy issues are addressed in the United States (Katz & Bradley, 2013).

Many modern economic and social gains are linked, in some form or another, to a particular technological platform: high-speed Internet access. Broadband networks are the foundation of much technological progress—they lie at the heart of a vibrantly inventive ecosystem of innovation and support a growing universe of devices, applications, content, and services, all of which deliver enormous benefits to end-users (FCC, 2010). These platforms also hold the potential for profoundly impacting the lives of those who possess the requisite skills to put their connections to productive and meaningful uses (ibid.). In urban environments in particular, these networks are viewed as increasingly vital tools for governing more effectively and for serving as the nucleus of data-driven "smart cities" that can enhance a range of services for residents (Goldsmith & Crawford, 2014). The potential for leveraging advanced communications tools in urban environments is thus significant and has generated a cottage industry around the smart city concept. Many now agree that nearly every facet of local government—from administering social services to streamlining traffic flows—can be disrupted by the right mixture of technology and expertise (McKinsey, 2013, pp. 23–25).

The seemingly essential nature of digital infrastructure—to government, institutions, residents, and businesses—has led some to argue that it is too important to be left to the private sector to build and operate (Crawford, 2013, pp. 254–270). Accordingly, there have been numerous calls in the recent past for cities to deploy their own networks in order to preserve local autonomy and maintain control over such a vital component of modern governance (Davidson & Santorelli, 2014, pp. 14–18). Many other rationales have been put forward in support of what is variously called "municipal broadband," "community broadband," or "government-owned broadband networks" (GONs). These efforts typically involve a desire by local government to compete with incumbent private firms or spur economic development (Davidson & Santorelli, 2013).

There is little doubt that broadband is becoming an indispensable asset for cities, but, as discussed in this chapter, municipal provision of the service, for public or private use, often makes little sense, especially in densely populated urban areas that are already well served by multiple entities. The history of GONs in the United States is littered with costly failures and numerous other examples of municipal networks that failed to justify their very high price tags. Instead of risking scarce public resources on duplicative networks that tend to fail in competitive environments, this chapter highlights the myriad of ways in which stakeholders at the city level can bolster the availability and use of broadband in urban environments.

As an overview, on the supply side, cities can work to enhance broadband availability by:

- Revisiting and revising local laws and regulations impacting the deployment of broadband networks—zoning ordinances; franchising requirements; rights-of-way management; and siting rules. Modernizing these rules to reflect the current contours of how services are provided will unlock additional private investment in broadband.
- Working in communities to identify real areas of need—*i.e.*, truly unserved areas—and partnering with private providers to determine the best way of bringing service to these areas.
- Refraining from building a GON in light of the many risks involved and channeling any available funding to seeding public-private partnerships that will address availability issues more efficiently and effectively.

On the demand side, local stakeholders can enhance informed use of advanced communications tools by:

- Prioritizing digital literacy as a core feature of any local effort aimed at bolstering technology use.
- Gathering local information and data about pockets of under-use and partnering with expert organizations—nonprofits; anchor institutions; private companies—to target outreach and training efforts.
- Developing a comprehensive framework for prioritizing connectivity and delivering training services across the continuum of use—from early education through workforce development and older adult care—by, among other things, integrating demand side imperatives into social service delivery.

The New Urban Core: Broadband's Ubiquitous Presence in the Modern City

Advanced communications networks, including those capable of enabling high-speed Internet connectivity, have always been more prevalent in urban areas than in other parts of the country. High population density, the clustering of businesses with significant data capacity needs, and the presence of large swaths of higher income and well-educated consumers are among the many reasons that service providers have targeted cities when rolling out advanced services. Government reports from the mid-1990s confirm that, by and large, many of the earliest forms of broadband—ISDN, DSL, and cable—were deployed primarily in major cities and metropolitan areas (FCC, 1999). Perhaps not surprisingly, urban consumers were much more likely than rural counterparts to be online during this initial phase of the commercial Internet market (NTIA, 1995, 1998, 1999, 2000).

As broadband evolved and became an affordable replacement for dial-up Internet service, cities continued to serve as laboratories for experimentation vis-à-vis new service offerings and business models. Cable broadband service, for example, was initially most prevalent in densely populated urban areas (FCC, 2002); some of the first mobile broadband networks were deployed in major cities (Kim, 2010); and wireless broadband services like Wi-Fi quickly became staples of Internet access in many downtown areas (NTIA, 2008). Also during the mid-2000s, cities began to experiment with networks of their own, with many viewing Wi-Fi as a potentially affordable means of blanketing an urban area and competing with incumbent providers (Gillett, 2006). However, as discussed in more detail below, many of these endeavors ultimately failed (Davidson & Santorelli, 2014, pp. 14–16).

In the last decade, the urban broadband market has grown even more robust. The percentage of Americans living in urban areas without access to a fixed (i.e., wireline) broadband connection of at least 25 megabits per second (Mbps) decreased from 11 percent in 2012 to four percent in 2014 (FCC, 2016). Mobile broadband connections of at least 10 Mbps are also more widely available in cities than in rural areas (ibid.). Data included in the National Broadband Map (2016) provides a more detailed picture of the availability of high-speed Internet access in urban America: the vast majority of the population in these areas have access to numerous broadband connections at a range of speeds provided by an array of wireline and wireless carriers.

The ubiquity and robustness of broadband in urban parts of the country are made even more evident when contrasted with availability in rural areas. An urban/rural divide in Internet access has been evident in the United States since the 1990s (NTIA, 1995), and persists to this day (CEA, 2015). The FCC (2016) reports that 39 percent of Americans in rural areas lack access to fixed broadband connections with speeds of at least 25 Mbps. This situation is somewhat less dire when a broader range of speeds and platforms is considered (National Broadband Map, 2016), but, even then, it is clear that rural areas remain among the most challenging to serve.

Connectivity challenges do remain in urban areas, most notably on the demand side. Despite ready access to a broadband connection, subscription rates in certain urban communities lag far behind the general population. These "digital divides" have been observed since the advent of mass market Internet access (NTIA, 1998) and persist in the broadband age (CEA, 2015; FCC, 2016; File & Ryan, 2014). Although numerous forces impact whether or not a particular person opts to purchase a broadband connection, demographic and socioeconomic factors, including income and education level, tend to play outsize roles (Davidson, Santorelli & Kamber, 2012; Horrigan, 2010; Horrigan & Duggan, 2015). Consequently, the urban poor and large percentages of African Americans and Hispanics living in cities tend to adopt broadband at much lower rates (FCC, 2016; Tomer & Kane, 2015).

The Failed Municipal Wi-Fi Experiment

Despite such robust availability and obvious needs on the demand side, responses to urban broadband challenges have revolved primarily around calls for even more supply of high-speed networks. It is within this context that advocacy for municipal broadband has become more pointed in recent years. But calls for municipal intervention into local broadband markets, even those that are well served, is not new. On the contrary, debates over whether and when a city should or should not build its own network have been part of the national broadband dialogue for many years (Santorelli, 2007).

Many current rationales for GONs are variations on themes, theories, and arguments put forward by some who expressed deep pessimism about the state of broadband in the United States in the early and mid-2000s. Despite significant gains across a number of important metrics—e.g., speed; pricing; and availability—there were some who viewed these advances and the policies enabling them as broadening, rather than shrinking, the digital divide (Feld, Rose, Cooper, & Scott, 2005). In the absence of drastic policy reversals, these advocates argued that government-owned and -operated broadband was the best way to inject competition into the broadband marketplace, nudge incumbents to lower prices, and assure universal availability (Davidson & Santorelli, 2014).

The first major wave of municipal broadband crested in the mid-2000s with the emergence of Wi-Fi. Many viewed this wireless technology as a potential game-changer vis-à-vis enhancing connectivity because it was relatively cheap to deploy; was amenable to mesh networking strategies, which could, in theory, bolster coverage; and was built on freely available unlicensed portions of the wireless spectrum. Indeed, for some Wi-Fi held the theoretical promise of "turn[ing] the airwaves into a commons without tragedy, and turn[ing] the economics of wireless [and broadband provision generally] on its head" (Anderson, 2003). As such, a primary justification for pursuing municipal Wi-Fi at the time was that it would offer residents a more affordable—even free—means of going online. Implicit in this argument was a belief, notwithstanding data to the contrary, that broadband provided by the private sector was either out of reach financially or not physically accessible by those living in destitute urban areas (Travis, 2006).

Many subsequent Wi-Fi projects in large urban environments faltered due to a combination of lackluster usage among residents, technological shortcomings, and budgetary pressures. The paradigmatic example of a failed muni Wi-Fi system from this era was in Philadelphia. In 2005, it committed to building a citywide Wi-Fi network in an effort to blanket the city with cheap Internet access and become the "next great city" by, among other things, offering the 55 percent of residents who were not online a convenient onramp (Lee, 2008). Despite much positive press about the potential for this municipal network to transform Philadelphia, the Wi-Fi system eventually collapsed after years of delays, technical difficulties, cost-overruns, and lackluster usage (ibid.; Santorelli, 2007). Indeed, in a city of 1.5 million people, less than 6,000 used the service (Lee, 2008).

The fallout from Philadelphia foreshadowed troubles in other urban areas. Between 2005 and 2008, a number of large cities—including Orlando, FL; Chicago, IL; Houston, TX; San Francisco, CA; St. Louis, MO; and Cincinnati, OH, among others—terminated or halted their municipal wireless plans (Keen, 2007).

These failures cast a chill over municipal Wi-Fi: between 2007 and 2010 the total number of cities that had deployed or were considering deploying public Wi-Fi systems decreased from over 400 to fewer than 200 (Davidson & Santorelli, p. 17).

The False Promise of Municipal Fiber

Notwithstanding these many high-profile failures and analyses suggesting that municipal broadband rarely made economic sense, advocacy for local government involvement in this space persisted. Once it was clear that Wi-Fi was not a viable platform for a GON, the focus shifted to higher capacity wireline alternatives like fiber-optics. The prevailing rationale for the focus on fiber was that it was "future-proof" because it could offer seemingly limitless capacity, in the form of gigabit connectivity, and would serve as a lure for new businesses, a generator of new jobs, a spur for start-ups, and the foundation on which the future economy would be built (NATOA, 2008). In short, geography would no longer matter if a city had a gigabit fiber network. Moreover, this embrace of fiber framed GONs as a kind of *fait accompli* because very few ISPs at the time planned to offer gigabit speeds in the near future because there was little, if any, demand for such high-speed connectivity (Atkinson, Schultz, Korte, & Krompinger, 2011; FCC, 2011).

The track record of fiber-based GONs that have been deployed in the United States over the last two decades is decidedly mixed. Few municipal systems have emerged as enduring successes. Indeed, many of the highest-profile municipal fiber networks that have been built in recent years have failed to live up to the hype that initially surrounded their launch. Moreover, those framed as successes tend to be outliers that arose from unique circumstances that are difficult to replicate. The ultimate demise of a much-touted early exemplar of a successful fiber-based GON deployed in Vermont's largest city—Burlington—offers an illustrative case study of some of the pitfalls that tend to arise.

In 2005, Burlington began offering residents and businesses a proprietary broadband service via a fiber-optic network that had initially been deployed for the exclusive use of city agencies. After securing tens of millions of dollars in financing, the network appeared to be on a path toward sustainability in 2007, but despite a positive cash flow and a slowly expanding subscriber base, by 2008 overall revenues were insufficient to cover its debt payments (Mitchell, 2011). By 2009, the municipal system had amassed significant debt, leading the city council to conclude that the system was "too deeply indebted to break even given the size of its customer base" (ibid., p. 4). Volatile finances and disputes with creditors led to a credit downgrade for the city, which led the city to explore selling the system (Niles, 2014).

Significant debt and meager demand have undermined many other fiber-optic GONs in the United States over the last few years. Two of the largest failed municipal fiber initiatives are located in Utah. Provo, the state's third largest city and an emerging hub for high-tech innovation, amassed about $60 million in debt to build a fiber-to-the-home (FTTH) network to residents in the early 2000s (Davidson & Santorelli, 2014, pp. 83–84). However, it struggled to build up a user base that would allow it to generate revenues sufficient to pay down its debt and to operate the network (ibid.). By 2008, the year in which city officials initially projected that the system would be profitable, Provo was forced to prop up the system with millions of dollars in subsidies borrowed from the city's coffers (Dana, 2008). After several more years of losses and under-use, the city had little choice but to sell the failing network to Google, which agreed to buy the system for $1, leaving taxpayers with tens of millions of dollars in debt to pay off over the next decade (Horiuchi, 2013).

In more remote parts of Utah, a consortium of smaller cities in the state banded together to launch the Utah Telecommunications Open Infrastructure Agency (UTOPIA), an ambitious effort to deploy a fiber network to connect residents in 16 municipalities (Davidson & Santorelli, 2014, p. 75). Construction began in 2003; the first phase of the network went live in 2005 (Schaff, 2012). Shortly thereafter, though, the multi-city system began to falter: over the next six years, the GON would amass hundreds of millions of dollars in debt as it struggled to attract the necessary number of subscribers needed to cover costs (ibid.). By 2013, the cities were forced to begin exploring ways in which it could leverage private-sector investments in order to keep the failing system afloat (Davidson & Santorelli, 2014, p. 77).

An alternative fiber-optic GON deployment model that has proven increasingly popular in mid-sized cities in the United States involves the use of municipal electric utilities to build and operate networks. The leading example often cited by GON advocates is the utility-built FTTH GON in Chattanooga, Tennessee. The fiber system deployed by the utility there, EPB, was funded in large part by a federal stimulus award in the amount of $111 million—a grant that was nominally targeted at supporting a nascent fiber-enabled "smart grid" to enhance electricity distribution—as well as loans made from its electric division to a newly formed fiber division (Davidson & Santorelli, 2015). Even so, the system still required nearly $200 million in additional debt financing (ibid.). However, the initial infusion of federal funding, coupled with the intra-utility loans underwritten by the rates paid by captive electric customers in the city, hastened both the construction of the GON and the ability of the system to generate a positive cash-flow in a much shorter timeframe than if it had been built just with debt financing (ibid.).

Even though Tennessee state law prohibits explicit cross-subsidization of municipal broadband by parent utilities, the close relationship between the EPB electric and fiber-optic divisions conveys a number of advantages to the utility—and to other municipal utilities that employ a similar model. In the case of EPB, investment in fiber for the ostensible purpose of bolstering its smart grid can be socialized amongst its captive electric ratepayers (i.e., every resident and business in its service territory). But in practice, this means that at least some percentage of the GON and its commercial broadband service is being subsidized by the parent utility (Ford, 2015). The benefits flow the other way as well. For example, EPB readily admits that "revenues from [its] Fiber Optics division have allowed the utility to defer rate increases that would have totaled 5 [percent] over the last four years" (EPB, 2015a). Nevertheless, EPB raised its rates in 2015 to "cover increases in operating costs" (EPB, 2015b). Whether and how these increased rates might be used for the benefit of the utility's broadband network remains to be seen.

Larger cities have mostly avoided being swept up in this second wave of municipal broadband. Indeed, the vast majority of municipalities that have pursued fiber-based GONs are small and mid-sized cities. Chattanooga, with a population of about 140,000, is among the largest cities to deploy a GON of any kind. Some large cities, though, have investigated the feasibility of deploying citywide fiber networks, either on their own or in partnership with a private provider. Recent examples have included formal and informal inquiries in cities like Baltimore (Dance, 2015), Chicago (Hinz, 2014), and Los Angeles (Buckley, 2015) regarding the viability of collaborating with private sector partners to enhance broadband offerings. A formal feasibility study conducted for Seattle recommended against a citywide GON, citing cost concerns and skepticism about whether the network would be able to sustain itself in the face of competition from incumbent firms (Beekman, 2015).

Ultimately, the seeming viability of the GON in Chattanooga and other cities that have benefited from unique circumstances, which has sparked interest in municipal broadband in cities large and small, is belied by lingering questions about whether these costly networks have delivered the many benefits that were initially promised by those seeking approval for the system. Recent studies have shown that the economic gains attributed to a particular municipal network were rarely the result of the type of straightforward cause-and-effect depicted by advocates—i.e., that the mere presence of the network led to specific economic benefits (Ford, 2015). On the contrary, most benefits, to the extent that any manifested, tended to result from numerous other, non-technological factors (e.g., traditional economic incentives to relocate or launch a business) that, together, subordinate the role the network played in realizing these gains. Indeed, leaving anecdotal evidence aside, there is little hard data to demonstrate that the GONs in cities like Chattanooga led to significant gains in job or high-tech business creation (Davidson & Santorelli, 2014, pp. 97–99).

The absence of empirical data supporting these kinds of claims raises important questions about the opportunity costs associated with a decision to pursue a GON, namely whether the money spent on the network could have been better spent elsewhere. In the context of working to improve broadband connectivity, an essential inquiry by local policymakers weighing a GON proposal is whether public funding could be more wisely invested in either forging a public-private partnership to bolster local broadband infrastructure or supporting targeted demand side activities in an effort to increase adoption rates.

The Political Economy of the Modern GONs Movement

Municipal broadband remains exceedingly rare in the United States. Including municipalities that have partially built or unused public broadband assets (i.e., dark fiber), as well as those with FTTH systems, one estimate from October 2015 pegs the total number of community fiber systems at 383 (Community Network Map, 2016). Of these, only 83 communities have "publicly owned FTTH network[s] reaching most or all of the community"; 50 of those communities, spread across 19 states, have "publicly owned network[s] offering at least 1 Gigabit services" (ibid.). To put this into perspective, there are over 19,000 municipalities in the United States (Hogue, 2013). Nevertheless, GONs have become a prominent part of larger discussions about broadband in the country because of efforts by the federal government to promote it as a means of enhancing competition and innovation in the U.S. market.

Over the last few years, federal policymakers have launched a series of initiatives aimed at supporting GONs, creating a very unique federal-local dynamic. Initially, these efforts revolved around channeling funds to support communities that wished to build middle-mile broadband networks (ASR Analytics, 2014). However, subsequent activities expanded to provide federal funding, advocacy support, and other resources for municipal provision of last-mile Internet connections like FTTH. These have included a series of ongoing meetings and programs arranged by the Department of Commerce aimed at working directly with "communities that want to expand their broadband capacity" (BroadbandUSA, 2016). Subsequently, President Obama expressed support for municipal broadband efforts and chided states for adopting laws impacting the deployment of these systems (White House, 2015). Shortly thereafter, the FCC acted to preempt two of these laws (FCC, 2015). Each of these activities has been justified as necessary to realizing national imperatives around ensuring that every American has "affordable and competitive broadband choices" (ibid.; Executive Office of the President, 2015; White House, 2015).

Although not entirely without precedent (Davidson, 2007), such direct coordination between the federal government and municipalities in this context is unusual because it ultimately undermines the ability of state governments to oversee their subdivisions. Indeed, by attempting to frame GONs as essential inputs to long-term economic prosperity in the United States and to realizing a subjective level of effective competition among ISPs, proponents have often sought to marginalize the role of state-level officials, particularly state legislatures, in these discussions. As a result, efforts by state legislatures to mediate the exploration of high-risk municipal broadband projects, typically via legislation to govern the process by which these networks are approved and built, are often dismissed out of hand as intrusive encroachments of municipal authority and cited as proof in favor of federal intervention (Sylvain, 2012).

From the perspective of bolstering broadband connectivity—a widely shared goal—the modern dynamic of federal-local cooperation and state-level marginalization is suboptimal for several reasons. First, it dismisses the profound importance of preserving a state's constitutionally-protected right to monitor its municipalities. In the GONs context, the need for robust oversight is especially acute because states maintain ultimate responsibility for the financial health of the cities and towns in their borders. Municipal broadband is an expensive undertaking, costing anywhere from a few million dollars to several hundreds of millions of dollars. When a network falters, local governments are forced to step in and prop up a failing system with funding and other support. Failing systems often result in credit downgrades, which increase borrowing costs and strain local finances. State policymakers thus have a compelling interest in adopting legislative safeguards to protect against these kinds of outcomes (State Associations, 2015).

Second, attempts to work around the states in this context have resulted in costly and unproductive litigation. That efforts to preempt or otherwise undermine state authority over GONs have proceeded in the first place is surprising given the clarity and consistency with which courts have ruled on these issues in the past; the Supreme Court has been firm in preserving the fundamental relationship between states and cities (Hunter v. City of Pittsburgh, 1907, p. 178). The Supreme Court has been equally unequivocal in the specific context of municipal broadband. In 2004, for example, the Court upheld a Missouri law that prohibited municipalities from offering telecommunications services (Nixon v. Missouri Municipal League, 2004). As such, it is unlikely that federal action to preempt or otherwise diminish state authority vis-à-vis GONs will withstand judicial scrutiny.

Finally, as explored more fully in the next section, states and state policymakers have important roles to play in the broadband space, both alone and in partnership with the municipalities they oversee. Circumventing the states in an effort to work directly with cities and towns is an inefficient—and likely illegal—approach to addressing pressing issues that require "all hands on deck" strategies.

How to Enhance Broadband Connectivity in Urban Areas

Though it implicates critical issues regarding the ability and extent to which municipalities can and should influence broadband deployment, the modern debate over GONs tends to be overly reductive because it presents only a binary choice between full-scale government involvement—in the form of a GON—and none at all. This Manichean worldview obscures the significant middle ground wherein local government, in partnership with counterparts at the state level and the private and nonprofit sectors, can successfully deploy effective strategies for addressing clearly defined broadband connectivity issues. These opportunities are especially ripe in urban areas due to the vibrancy and diversity of resources at the disposal of city policymakers.

Bolstering the Supply of Broadband

There are numerous opportunities for cities to leverage their assets in an effort to enhance the private provision of broadband services. Many of these involve actions that would greatly enhance new network deployment by streamlining access to key infrastructural inputs. In general, municipalities maintain significant authority over how firms access and use the physical infrastructure over which they will build networks. These activities include right-of-way management (e.g., via franchising) and the range of rules and regulations impacting access to other parts of a city's assets (like granting approval to construct a wireless tower or adopting zoning laws impacting where certain network equipment can be placed). Properly managed, cities can take advantage of these assets, and the natural economies of scale, scope, and density evident in most urban areas, to bolster the supply of broadband.

An important first step is determining areas of real need and then tailoring strategies to address those specific issues. In growth cities like New York and Chicago, broadband is difficult to access in areas that were originally zoned for commercial use. In many instances, areas that were formerly industrial zones are quickly transforming into hubs for start-up businesses and workers. These areas tend to fall outside of the historical service footprint of ISPs, leaving new tenants without ready access to broadband. To address these kinds of issues in an effective and timely manner, cities have begun working with stakeholders in new neighborhoods and broadband service providers to determine how best to serve the needs of these communities (Davidson & Santorelli, 2014).

Beyond these discrete access issues, cities also have significant latitude to adjust the ways in which it leverages its rights-of-way in support of network construction. The Google Fiber experiment is instructive. Google only works with cities that are willing to accede to its terms for deploying its FTTH system. The agreement that was eventually worked out with its first partner city, Kansas City, differed in many ways from the franchise agreements the municipality historically entered into with incumbent ISPs (ibid., pp. 121–124). In particular, the agreement included very few binding commitments for Google, but it enshrined a number made by the city regarding granting the company priority—and, in many cases, free—access to rights-of-way and other key inputs. The city also agreed to Google's unique demand-driven deployment model, one that prioritizes deployment to areas with the greatest demand for its service but risks omitting communities, like low-income and minority households, that have historically under-adopted these services (ibid.).

Google has since used this model in a number of other cities, including Austin and Provo. Recognizing that a level playing field among competitors is essential to spurring investment, fostering sustainable competition, and encouraging further network deployment, officials in Google partner cities and elsewhere have begun to make these alternative arrangements available to all ISPs. Verizon, for example, worked with officials in Boston to develop a similar model for deploying its FTTH system (Woodward & Chesto, 2016). More formal franchise reform efforts—at the city and state levels—could follow in light of the successes and failures of this unique model.

Cities are also experimenting with new ways of leveraging underused assets like telephone booths to support broadband deployment. New York City, for example, recently decided to convert 7,500 telephone kiosks into Wi-Fi hotspots over the next few years (Flegenheimer, 2014). Unlike during the initial Wi-Fi push in cities a decade ago, this initiative relies exclusively on private funding to deploy the network; the city will collect fees based on digital advertising revenue that the vendor will generate based on data it collects from users. This arrangement has raised significant privacy concerns, though, as some have noted that users will be largely unaware of the extent to which the hotspot is collecting and monetizing "potentially sensitive personal, locational and behavioral data" (Dean, 2016). Even so, many cities are examining how to inject intelligence—and real-time data collection and processing—into a broader array of objects like street lights, traffic lights, street furniture, and other components of the "smart city" (Clark, 2015).

State and federal government policymakers also have roles to play in helping bolster broadband supply at the local level. State legislatures and regulators, for example, could revisit older laws and policies, like those governing review or construction processes, which might be impacting deployment at the local level.

Similarly, federal policymakers have acted on several occasions in the recent past to streamline local review of wireless network deployment. These actions were deemed necessary in order to inject more consistency and clarity about how long localities can take to approve or deny applications for building a tower or erecting an antenna (FCC, 2009).

Improving Broadband Adoption and Digital Literacy

Cities are especially well positioned to address digital divide issues that have plagued urban areas for decades. As noted previously, issues relating to broadband adoption and digital literacy challenges are acutely evident in many urban communities, yet these issues have not received nearly the same level of attention or resources from city, state, or federal policymakers. Instead, the policy focus has remained almost exclusively on the supply of broadband, notwithstanding the systemic disparities in adoption rates across a range of user communities. Part of the reason for this is the implicit assumption that demand side issues will be resolved once there is ample supply of cheap or free broadband. Though appealing, this reductive cause-and-effect has been questioned by social scientists, researchers, practitioners, and others who have worked to identify and better understand the complex mechanics associated with broadband adoption across key demographics (Horrigan, 2010, 2014).

There is growing evidence that a more robust focus by communities on increasing broadband adoption, delivering training services, tying to social service delivery, and improving digital literacy skills yield significant and lasting economic and social gains (Davidson, Santorelli & Kamber, 2012; FCC, 2010). Because broadband adoption issues are best addressed at the hyper-local level, cities are in the best position to work with local stakeholders on these issues—a simple but powerful fact that local policymakers should embrace rather than undermine by continuing to focus only on supply issues. To these ends, public-private partnerships (PPPs) have proven especially effective in spurring broadband adoption and assuring productive uses of digital tools.

The most impactful PPPs are those that tap into local social infrastructures, the network of local institutions and organizations—e.g., community centers, libraries, schools, senior centers, churches, nonprofit organizations that specialize in digital literacy, and companies like ISPs, with roots in the municipality—that serve as ideal conduits for channeling education, outreach, and training programs in areas that need it most. This approach yields positive and sustainable results because these groups have succeeded in engendering high levels of trust with residents and have demonstrated an ability to deliver community-specific services, including key social services like workforce training, government benefits, education services, and healthcare information, among many others (Davidson, Santorelli & Kamber, 2012; Rhinesmith, 2016).

Cities that collaborate with expert entities to develop and implement comprehensive, community-specific adoption and training strategies are much more likely to begin closing digital divides and empowering more residents with the ability to use broadband to improve their lives. A necessary first step, of course, is for cities to acknowledge that demand side challenges exist and that solving them requires much more than doubling down on supply side efforts. Unfortunately, the focus of many cities, especially in large urban areas like New York, remains trained on the supply side, leaving important demand side issues to worsen.

Conclusion

That broadband is transforming urban life—for residents, government, businesses, and core institutions—is indisputable. Debate in this context tends to arise, though, when stakeholders attempt to identify and address challenges that are impeding realization of this technology's full potential. In the urban context, the data is clear: broadband is widely available but under-adopted in certain communities. Digital literacy also remains largely overlooked. To the extent a city wishes to invest scarce public resources in addressing broadband issues, it would be wise to avoid deploying its own network or otherwise over-committing on the supply side, and instead target well defined problems like working to ensure that every resident is aware of and able to use this powerful tool. At the same time, cities should reevaluate the many rules and regulations impacting the deployment of broadband infrastructure in an effort to encourage additional private investment and innovation at the network level. Approaching broadband in this way will assure that those urban areas that truly embrace the "smart city" concept will have a digitally ready and engaged user base.

References

Anderson, C. (2003, May). The Wi-Fi revolution. *Wired*. Retrieved from http://www.wired.com/2003/05/wifirevolution/

ASR Analytics. (2014, September). Final report: Social and economic impacts of the broadband technology opportunities program. *Report to the National Telecommunications and Information Administration*. Washington, DC. Retrieved from https://www.ntia.doc.gov/files/ntia/publications/asr_final_report.pdf

Atkinson, R., Schultz, I., Korte, T., & Krompinger, T. (2011, May). Broadband in America – 2nd edition. *Columbia Institute for Tele-Information*. Retrieved from http://issuu.com/romyciti/docs/complete_final_jun14-1/1?e=2686346/3479190

Beekman, D. (2015, June 10). Cost of municipal broadband for Seattle less than estimated. *Seattle Times*. Retrieved from http://www.seattletimes.com/seattle-news/politics/cost-of-municipal-broadband-for-seattle-less-than-estimated/

BroadbandUSA. (2016). About. *National Telecommunications and Information Administration, U.S. Department of Commerce*. Washington, DC. Retrieved from http://www2.ntia.doc.gov/about

Buckley, S. (2015, June 11). Los Angeles approves RFP for planned 1 Gbps muni network that could cost $3B–$5B. *Fierce Telecom*. Retrieved from http://www.fiercetelecom.com/story/los-angeles-approves-rfp-planned-1-gbps-muni-network-could-cost-3b-5b/2015-06-11

CEA (Council of Economic Advisors). (2015). Mapping the digital divide. *The White House*. Retrieved from https://www.whitehouse.gov/sites/default/files/wh_digital_divide_issue_brief.pdf

Clark, L. (2015, May 18). 'Street furniture' adapts to the needs of everyone. *Wired UK*. Retrieved from http://www.wired.co.uk/news/archive/2015-05/18/responsive-street-furniture-designs

Community Network Map. (2016). *Community Broadband Networks, Institute for Local Self-Reliance*. Retrieved from http://muninetworks.org/communitymap

Crawford, S. (2013). *Captive audience: The telecom industry and monopoly power in the new gilded age*. New Haven, CT: Yale University Press.

Dana, J. (2008, April 22). Provo eyes ways to fix its network. *Deseret News*. Retrieved from http://www.deseretnews.com/article/695272699/Provo-eyes-ways-to-fix-its-network.html?pg=all

Dance, S. (2015, October 18). Baltimore looks to join 'gigabit cities,' including Westminster, in boosting broadband Internet capacity. *Baltimore Sun*. Retrieved from http://www.baltimoresun.com/business/technology/bs-md-ci-broadband-expansion-20151013-story.html

Davidson, C., & Santorelli, M. (2013). Evaluating the rationales for government-owned broadband networks. *Advanced Communications Law & Policy Institute at New York Law School*. Retrieved from http://www.nyls.edu/advanced-communications-law-and-policy-institute/wp-content/uploads/sites/169/2013/08/Davidson-Santorelli-Evaluating-the-Rationales-for-GONs-March-2013.pdf

Davidson, C., & Santorelli, M. (2014). Understanding the debate over government-owned broadband networks: Context, lessons learned, and a way forward for policy makers. *Advanced Communications Law & Policy Institute at New York Law School*. Retrieved from http://www.nyls.edu/advanced-communications-law-and-policy-institute/wp-content/uploads/sites/169/2013/08/ACLP-Government-Owned-Broadband-Networks-FINAL-June-2014.pdf

Davidson, C., & Santorelli, M. (2015, September). Chattanooga case study (updated). *Advanced Communications Law & Policy Institute at New York Law School*. Retrieved from http://www.nyls.edu/advanced-communications-law-and-policy-institute/wp-content/uploads/sites/169/2013/08/ACLP-Chattanooga-Case-Study-updated-October-2015.pdf

Davidson, C., Santorelli, M., & Kamber, T. (2012). Toward a more inclusive definition of broadband adoption. *International Journal of Communication, 6*, 2255–2275.

Davidson, N. (2007). Cooperative localism: Federal-local collaboration in an era of state sovereignty. *Virginia Law Journal, 93*, 959–1034.

Dean, B. (2016, January 25). The heavy price we pay for 'free' Wi-Fi. *The Conversation*. Retrieved from https://theconversation.com/the-heavy-price-we-pay-for-free-wi-fi-52412

EPB. (2015a, May 22). EPB asks board for electric rate increase. *EPB*. Chattanooga, TN. Retrieved from https://www.epb.net/mediarelations/news/news_archive/epb-asks-board-for-electric-rate-increase/

EPB. (2015b, June 26). EPB board approves electric rate increase. *EPB*. Chattanooga, TN. Retrieved from https://www.epb.net/mediarelations/news/news_archive/epb-board-approves-electric-rate-increase

Executive Office of the President. (2015). Community-based broadband solutions: The benefits of competition and choice for community development and high speed internet access. *The White House*. Washington, DC. Retrieved from https://www.whitehouse.gov/sites/default/files/docs/community-based_broadband_report_by_executive_office_of_the_president.pdf

FCC (Federal Communications Commission). (1999). In the matter of inquiry concerning the deployment of advanced telecommunications capability to all Americans in a reasonable and timely fashion, and possible steps to accelerate such deployment pursuant to Section 706 of the telecommunications act of 1996, report. CC Docket No. 98–146. Washington, DC.

FCC (Federal Communications Commission). (2002). In the matter of inquiry concerning the deployment of advanced telecommunications capability to all Americans in a reasonable and timely fashion, and possible steps to accelerate such deployment pursuant to Section 706 of the telecommunications act of 1996, third report. CC Docket No. 98–146. Washington, DC.

FCC (Federal Communications Commission). (2009). In the matter of petition for declaratory ruling to clarify provisions of Section 332(c)(7)(B) to ensure timely siting review and to preempt under Section 253 state and local ordinances that classify all wireless siting proposals as requiring a variance, declaratory ruling. WT Docket No. 08–165. Washington, DC.

FCC (Federal Communications Commission). (2010). Connecting America: The national broadband plan. Washington, DC. Retrieved from https://transition.fcc.gov/national-broadband-plan/national-broadband-plan.pdf

FCC (Federal Communications Commission). (2011). Measuring broadband America. Washington, DC. Retrieved from https://transition.fcc.gov/cgb/measuringbroadbandreport/Measuring_U.S._-_Main_Report_Full.pdf

FCC (Federal Communications Commission). (2015). City of Wilson, North Carolina petition for preemption of North Carolina general statute sections 160A-340 et seq., the Electric Power Board of Chattanooga, Tennessee petition for preemption of a portion of Tennessee code annotated section 7-52-601, Memorandum Opinion and Order. 30 FCC Rcd 2408. Washington, DC.

FCC (Federal Communications Commission). (2016). In the matter of inquiry concerning the deployment of advanced telecommunications capability to all Americans in a reasonable and timely fashion, and possible steps to accelerate such deployment pursuant to Section 706 of the telecommunications act of 1996, as amended by the broadband data improvement act, 2016 broadband progress report. GN Docket No. 15–191. Washington, DC.

Feld, H., Rose, G., Cooper, M., & Scott, B. (2005, April). Connecting the public: The truth about municipal broadband. *Free Press*. Retrieved from http://www.freepress.net/sites/default/files/fp-legacy/mb_white_paper.pdf

File, T., & Ryan, C. (2014, November). Computer and internet use in the United States: 2013. *American Community Survey Reports, U.S. Census Bureau.* Retrieved from http://www.census.gov/content/dam/Census/library/publications/2014/acs/acs-28.pdf

Flegenheimer, M. (2014, November 17)."Pay phones in New York City will become free Wi-Fi hot spots. *New York Times.*

Ford, G. (2015, January 20). Why Chattanooga is not the "poster child" for municipal broadband. *Phoenix Center for Advanced Legal & Economic Public Policy Studies.* Retrieved from http://www.phoenix-center.org/perspectives/Perspective15-01Final.pdf

Gillett, S. (2006). Municipal wireless broadband: Hype or harbinger? *Southern California Law Review, 79,* 561–593.

Glaeser, E. (2011). *Triumph of the city: How our greatest invention makes us richer, smarter, greener, healthier, and happier.* New York, NY: Penguin.

Goldsmith, S., & Crawford, S. (2014). *The responsive city: Engaging communities through data-smart governance.* San Francisco, CA: Jossey-Bass.

Hinz, G. (2014, February 9). City offers big carrot for super-fast Internet. *Crain's Chicago Business.* Retrieved from http://www.chicagobusiness.com/article/20140209/BLOGS02/140209794/city-offers-big-carrot-for-super-fast-internet

Hogue, C. (2013, September 26). Government organization summary report: 2012. *U.S. Census Bureau.* Washington, DC. Retrieved from http://www2.census.gov/govs/cog/g12_org.pdf

Horiuchi, V. (2013, April 24). Council approves iProvo sale to Google. *Salt Lake Tribune.* Retrieved from http://www.sltrib.com/sltrib/news/56206589-78/google-network-fiber-provo.html

Horrigan, J. (2010). Broadband adoption and use in America. *OBI Working Paper Series No. 1,* Washington, DC: Federal Communications Commission. Retrieved from https://transition.fcc.gov/national-broadband-plan/broadband-adoption-in-america-paper.pdf

Horrigan, J. (2014, March). The essentials of connectivity. Retrieved from http://corporate.comcast.com/images/Final_IE_Research_Full_Paper.pdf

Horrigan, J., & Duggan, M. (2015). Home broadband 2015. Washington, DC: Pew Research Center. Retrieved from http://www.pewinternet.org/files/2015/12/Broadband-adoption-full.pdf

Hunter v. City of Pittsburgh. (1907). 207 U.S. 161.

Katz, B., & Bradley, J. (2013). *The metropolitan revolution: How cities and metros are fixing our broken politics and fragile economy.* Washington, DC: Brookings Institution Press.

Keen, J. (2007, September 20). Cities turning off plans for Wi-Fi. *USA Today.*

Kim, R. (2010, October 6). Verizon wireless to light up 38 cities with LTE in 2010. *GigaOm.* Retrieved from https://gigaom.com/2010/10/06/verizon-wireless-to-light-up-38-cities-with-lte-in-2010/

Lee, D. P. (2008, September 24). Power: Whiffing on Wi-Fi. *Philadelphia Magazine.* Retrieved from http://www.phillymag.com/articles/power-whiffing-on-wi-fi/

McKinsey & Co. (2013). How to make a city great. Retrieved from http://www.mckinsey.com/~/media/mckinsey/dotcom/insights/urbanization/how%20to%20make%20a%20city%20great/how_to_make_a_city_great.ashx

Mitchell, C. (2011, August). Learning from Burlington Telecom. *Institute for Local Self-Reliance*. Retrieved from http://www.muninetworks.org/sites/www.muninetworks.org/files/bt-lessons-learned.pdf

National Broadband Map. (2016). Broadband statistics report: Broadband availability in urban vs. rural areas. Washington, DC: U.S. Department of Commerce. Retrieved from http://www.broadbandmap.gov/download/Broadband%20Availability%20in%20Rural%20vs%20Urban%20Areas.pdf

NATOA. (2008). Broadband as an economic development tool. Retrieved from http://www.natoa.org/policy-advocacy/documents/NATOABroadbandEconStimulus.pdf

Niles, H. (2014, November 10). Burlington Telecom's interim sale approved. *VTDigger.org*. Retrieved from http://vtdigger.org/2014/11/10/burlington-telecoms-interim-sale-approved/?utm_source=rss&utm_medium=rss&utm_campaign=burlington-telecoms-interim-sale-approved

Nixon v. Missouri Municipal League. (2004). 541 U.S. 125.

NTIA (National Telecommunications and Information Administration). (1995). Falling through the Net: A survey of "have nots" in rural and urban America. Washington, DC: U.S. Department of Commerce. Retrieved from http://www.ntia.doc.gov/ntiahome/fallingthru.html

NTIA (National Telecommunications and Information Administration). (1998). Falling through the Net II: New data on the digital divide. Washington, DC: U.S. Department of Commerce. Retrieved from http://www.ntia.doc.gov/report/1998/falling-through-net-ii-new-data-digital-divide

NTIA (National Telecommunications and Information Administration). (1999). Falling through the Net III: Defining the digital divide. Washington, DC: U.S. Department of Commerce. Retrieved from http://www.ntia.doc.gov/legacy/ntiahome/fttn99/FTTN.pdf

NTIA (National Telecommunications and Information Administration). (2000). "Falling through the Net IV: Toward digital inclusion. Washington, DC: U.S. Department of Commerce. Retrieved from http://www.ntia.doc.gov/files/ntia/publications/fttn00.pdf

NTIA (National Telecommunications and Information Administration). (2008). Networked nation: Broadband in America in 2007. Washington, DC: U.S. Department of Commerce. Retrieved from http://www.ntia.doc.gov/files/ntia/publications/networkednationbroadbandinamerica2007_0.pdf

Rhinesmith, C. (2016). Digital inclusion and meaningful broadband adoption initiatives. *Benton Foundation*. Retrieved from https://www.benton.org/sites/default/files/broadbandinclusion.pdf

Santorelli, M. (2007). Rationalizing the debate over municipal broadband. I/S: A Journal of Law & Policy for the Information Society, *3*, 43–82.

Schaff, J. (2012). Report to the Utah legislature: A performance audit Utah telecommunication open infrastructure agency. *Office of the Legislative Auditor General.* Salt Lake City, UT. Retrieved from http://le.utah.gov/audit/12_08rpt.pdf

State Associations. (2015, September 18). Brief for the National Governors Association, National Conference of State Legislatures, and Council of State Governments as Amici Curiae in

Support of Petitioner and Urging Reversal. State of Tennessee v. FCC, No. 15–3291, Sixth Circuit Court of Appeals. Retrieved from http://muninetworks.org/sites/www.muninetworks.org/files/Natl-Gov-Assn-NCSL-CSG-Amicus-Con%20Local-Choice-copy.pdf

Sylvain, O. (2012). Broadband localism. *Ohio State Law Journal, 73*, 796–838.

Tomer, A., & Kane, J. (2015). *Broadband adoption rates and gapes in U.S. metropolitan areas.* Washington, DC: Brookings Institution. Retrieved from http://www.brookings.edu/~/media/research/files/reports/2015/12/07-broadband/broadband-tomer-kane-12315.pdf

Townsend, A. M. (2013). *Smart cities: Big data, civic hackers, and the quest for a new utopia.* New York, NY: W. W. Norton & Co.

Travis, H. (2006). Wi-Fi Everywhere: Universal broadband access as antitrust and telecommunications policy. *American University Law Review, 55*, 1697–1800.

White House. (2015, January 13). Fact sheet: Broadband that works: Promoting competition & local choice in next-generation connectivity. *Office of the Press Secretary.* Washington, DC. Retrieved from https://www.whitehouse.gov/the-press-office/2015/01/13/fact-sheet-broadband-works-promoting-competition-local-choice-next-gener

Woodward, C., & Chesto, J. (2016, April 12). FiOS rollout in Boston could take up to 6 years. *Boston Globe.* Retrieved from https://www.bostonglobe.com/business/2016/04/12/verizon-fios-finally-coming-boston-mayor-announces-fiber-network/xv53GVErnPcGMOCokMIHiO/story.html

CHAPTER ELEVEN

Broadband Adoption and Access in New York City

A Case Study

EMILY M. LONG VITO
Assistant Director of Development, Film Society of Lincoln Center

As far as American urban landscapes go, one can hardly find a space with more communication needs than New York City. Home to over 8.5 million residents, the nation's most populous public school system, more Fortune 500 companies than any other state and an expanding technology sector (71% growth over the last ten years), the people who live and work in New York are part of a dynamic environment with urgent needs for connectivity and access (Forman, 2015; Hackett, 2015).

New York also happens to be at a point of inflection in its efforts to meet those needs. Current Mayor Bill de Blasio was sworn into office in 2013 on a mandate to address growing rates of inequality; his "tale of two cities" campaign drew attention to immense disparities in income, health and education between the city's top wage earners and those living in poverty. Increasing availability and adoption of broadband Internet has been one strand of his strategy to address these issues, and the Mayor's Office of Technology and Innovation was established to centralize and manage broadband expansion and adoption. The de Blasio administration's efforts coincide with the period of time following the end of the federal Broadband Technology Opportunities Program (BTOP), in which the City of New York was awarded nearly $14 million to upgrade and expand public computer centers and establish a range of digital literacy training programs to marginalized communities citywide (United States Department of Commerce National Telecommunications & Information Administration, 2015).

Now the city is halfway through the first term of its first new mayor since Michael Bloomberg's remarkable twelve years in the role. Millions of BTOP dollars have been spent. Technology continues its march toward developing devices reliant on fast Internet speeds, and Americans continue to buy those devices and use them for constant connectivity for business, education, entertainment and socializing. Now, to paraphrase a former populist mayor, how is New York City doing?

Defining Access and Adoption

Before proceeding with a discussion of New York City's efforts to increase broadband access and adoption, it is important to first review what is meant by the terms "access" and "adoption."

Of the two terms, access is the least problematic. When asked to define broadband access, experts agreed it is achieved when one can connect to services running at speeds defined by the Federal Communications Commission (FCC) as advanced. In its 2015 broadband progress report released in February 2015, the FCC determined speeds of 25 Mpbs for downloading and 3 Mpbs for uploading as meeting needs for advanced telecommunications capability – in comparison, advanced speeds in 2010 were considered those of at least 4 Mbps/1 Mbps (Federal Communications Commission, 2015, p. 29).

However, it should be noted that accessibility differs from availability: just because a service is technically available does not mean it is accessible, as in cases where the available service is cost-prohibitive for local residents, or if the available service is unreliable. For purposes of this chapter, broadband access refers to service that is affordable to residents in the immediate area and which reliably meets the FCC's current standards for advanced telecommunications capability.

Defining adoption is more difficult – indeed, some of the people interviewed for this chapter eschewed setting a universal definition at all, either because it is constantly changing or because it is dependent upon community needs. Adoption needs to be broken out in two forms: the ability of a person to physically use a given technology (thus dependant upon physical access), and meaningful adoption, in which a person uses a technology in a way that is meaningful for them. For example, community members using broadband for socializing or gaming may believe themselves to be meaningful adopters, if this is their defined purpose for technology. However, members of a different community may view adoption as related to using technology to access government services, educational opportunities or other online resources. This can become a sticking point when public funds are used to facilitate broadband access for a community whose definition of meaningful adoption may differ from another's. This chapter will use the framework developed by the Open Technology Institute (OTI), in which the term "meaningful adoption" accounts for historical and contextual factors impacting digital choices.

One reason the OTI framework is useful is because it developed from looking at community-based technology needs and outcomes in the BTOP programs in Philadelphia and Detroit. "Usually adoption is measured by number of [home] subscriptions," said Greta Byrum, Senior Analyst in the Resilient Communities Program at the New America Foundation, who also worked on the OTI project. "But we noticed a lot of new users wanted to be surrounded by community, so they would go to public spaces. We also noticed a spectrum of adoption. When the push for adoption happened as part of supported community, the outcomes were better. We engaged in a collaborative evaluation process and created a set of shared goals around civic engagement, employment. We found that when BTOP resources were embedded in community centers, those were places people were going anyway, and that they trusted (G. Byrum, personal communication, December 17, 2015)."

Thus, for the purposes of this chapter, use of the term "adoption" refers only to regular use of a technology and makes no inference about how that technology is being utilized.

Barriers to Broadband Adoption and Access

With regard to adopting and accessing broadband, the barriers faced by New Yorkers are similar to those faced by citizens nationwide. According to a recent study from the Pew Research Center, 33% of Americans who do not have broadband access cite monthly subscription costs as the primary reason for not having access at home; an additional 10% said owning a computer is too expensive, making for a total 43% of Americans lacking broadband due to cost (Horrigan & Duggan, 2015). This persists even as 65% of non-adopters agree that lacking access is a major disadvantage for them when it comes to seeking job opportunities, accessing government services and learning new things in general (48% agreed in 2010). The same study finds home broadband adoption has decreased slightly from 70% in 2013 to 67% in 2015, suggesting that perhaps costs are or are perceived to be increasing.

Although a comprehensive study has yet to be conducted for New York City on the same scale as the national study conducted by Pew, most of the experts I spoke with in researching this piece agreed that cost is the largest barrier for increasing broadband home access. Indeed, as of the spring of 2014, 23% of New York City residents were living under the government-determined supplemental poverty line, and 37% suffered "severe material hardship (Wimer et al., 2014)."

Other contributing reasons cited included a lack of understanding benefits of home broadband access, poor customer service and a lack of service availability.

One unique factor affecting broadband access and adoption issues in New York is the sheer size of its population and number of stakeholder groups. "The city of Boston has fewer residents than New York City has schoolchildren," said Deb Socia, Executive Director of Next Century Cities, a nonprofit organization working to expand reliable, fast and affordable Internet access in communities nationwide (D. Socia, personal communication, December 15, 2015). "When I worked in Boston, I knew most of the nonprofit leaders and it was possible to get all of them into one conference room. That's not possible for New York, so the creative solutions here are pretty extraordinary."

Socia went on: "NYC has an incredibly complex problem. What I like is they haven't said there's one solution. The other thing that's really exciting is how engaged city government is, which is not always the case, especially around the issue of adoption."

Efforts from the Mayor's Office of Technology and Innovation and the Department of Information Technology and Telecommunications (DoITT)

As mentioned earlier, Mayor de Blasio established the Mayor's Office of Technology and Innovation to centralize the city's broadband expansion and adoption strategic efforts. This strategy is outlined in the city's "Smart City, Equitable City" plan, which includes expanding the public's access to technology as well as the learning and workforce opportunities related to technology. Making broadband more accessible is key to these efforts, and leading this charge is Maya Wiley, Counsel to the Mayor, whose team works closely with the Office of Technology and Innovation as well as the Department of Information Technology and Telecommunications (DoITT).

Under the de Blasio administration, the drive to expand broadband access is directly related to economic divisions. "It's not broadband for its own sake," said Josh Breitbart, Special Advisor for Broadband in the Mayor's Office of Technology and Innovation. "It's about broadband as a lever to address economic inequities (J. Breitbart, personal communication, October 27, 2015)." This outlook has directly impacted strategy for increasing broadband access and adoption in the city. This strategy is carefully laid out in the Mayor's OneNYC plan, together with the promise that, "Every resident and business will have access to affordable, reliable and high-speed broadband service everywhere by 2025." The plan goes on to say that 22% of New Yorkers do not have Internet service at home, and that commercial high-speed connectivity options are outside the affordability of many small businesses (de Blasio & Shorris, 2015).

This is not the first time the City has tried for full broadband access. An attempt to connect 100% of New Yorkers with competitively priced broadband was made in 2008, when the City entered a franchise agreement with Verizon in which the service provider agreed to make its FiOS service available to all New York City residents by June 30, 2014. However, in June of 2015, DoITT released an audit of the Verizon FiOS implementation asserting that Verizon had not met its obligation to the city (de Blasio & Roest, 2015). DoITT also stated that they were unable to complete the audit as per Generally Accepted Government Auditing Standards (GAGAS) due to a lack of cooperation from Verizon in providing complete and accurate reports. Months later, in October, a City Council hearing was convened at which Verizon officials reiterated in writing their claim that Verizon "takes great pride in being the first cable operator ever to make its services available *citywide* – in all five boroughs – and thus to provide *all* of the City's residents with a competitive choice for cable television services, delivered over the only all-fiber network in the City (Verizon, October 2015)."

In reviewing the hearing documents, it seems that the crux of this disagreement lay in different understandings of the term "passed": for purposes of this implementation, Verizon defines a residence has been "passed" if it made at least one attempt to provide service to the building, and the City defines "passed" as service actually being available. As Verizon representatives themselves pointed out during the hearing, language defining "passed" is not included in the original franchise agreement. Verizon explains that such language does exist in the City's 2011 agreements with its competitors Cablevision and Time Warner Cable, in which it is stated that "a household is 'passed' when functioning System facilities have been installed in the street fronting the building in which the household is located (Verizon, June 2015)." Verizon goes on to say that this point was discussed during negotiations with the City, but not included in contracts because the service provider had concerns about the phrase "in the street fronting the building". As testimony documents return again and again to the issue over the definition of passage, it seems that this loophole in language might be the largest reason that this landmark agreement did not ultimately result in high-speed Internet service availability for 100% of New York City residents and businesses as intended by June 2014.

Holding business franchises accountable for their commitments is part of the strategy coming from the Mayor's Office of Technology and Innovation. In addition, New York City is a participating community in the United States Department of Housing and Urban Development (HUD) pilot ConnectHome program, which seeks to accelerate broadband adoption and access for families living in HUD-assisted living nationwide. In July 2015, Mayor de Blasio and Counsel to the Mayor Maya Wiley announced a $10 million pilot program bringing free broadband to over 16,000 New Yorkers living in public housing developments in Queens, Brooklyn and the Bronx (Office of the Mayor, 2015). As noted earlier, Wiley leads the Office of Technology and Innovation, which helms the execution of this pilot program.

Another major effort is LinkNYC, a DoITT broadband access program made possible through a franchise agreement with CityBridge (CityBridge is technically a consortium of companies; one now known as Intersection is funded by Google's Sidewalk Labs [Silbey 2015]. This may explain why the kiosks will feature touch-screen Android tablets). The program essentially turns out-of-use public payphone kiosks into free public Wi-Fi access spots, or Links, across all five boroughs of the city; access is provided within 150 feet of a Link kiosk (Robbins, 2016). LinkNYC refers to the network itself that will consist of up to 10,000 Links, to be built and maintained by CityBridge. Advertising messages on the Link kiosks are expected to generate $500 million in revenue for the City. In addition to providing free Wi-Fi access, Links will also allow mobile device charging and free phone calls to be made anywhere in the United States. This program is seen as a big win for the Mayor's Office of Technology Innovation; as Josh Breitbart said, "LinkNYC is a huge milestone, going from payphone infrastructure to an aggressive plan to vastly increase coverage and speed."

Free and "limited free" Wi-Fi is also available in New York City parks locations through a partnership struck by DoITT and the Parks Department with Time Warner Cable and Cablevision. Known as Parks Wi-Fi, the program was launched in 2013 with current maintenance plans to run through 2020 (New York City Department of Parks and Recreation, 2013).

Both of these programs are considered public-private partnerships, in which a privately owned company partners with a public entity, such as the City of New York, to provide a public service – in this case, free Wi-Fi access (while neither provide fixed broadband access, they are nonetheless listed among of DoITT's broadband access initiatives). Partnerships such as these have raised concerns for some community advocates, who question whether these programs can truly be considered free if service is contingent on a consumer's willingness to supply personal data or be marketed to while using the service.

CityBridge's privacy policy says it will not sell user data, but it is able to use data it collects to deliver more targeted ads. In addition to user data, the kiosks themselves are also "smart," allowing them to use current weather conditions, time of day and other local businesses to determine what types of ads pedestrians might see. If a person logs on to kiosks using a phone that has a mobile applications with location services enabled, movement throughout the city can also be tracked to an extent. The Links will also exponentially extend the number of opportunities advertisers have to apply what they learn from user data – the kiosks are estimated to provide more than 9 million square inches of new advertising space throughout the city (Zara, 2014). Notably, the kiosks will be spread throughout the five boroughs of New York City, and not, as some advocates feared, solely limited to high traffic areas in wealthier or commercial parts of Manhattan. Manhattan will, however, receive the bulk of the kiosks – 52%, with the rest divided between Queens, Brooklyn, The Bronx and Staten Island (Robbins, 2016).

The office of Gale Brewer, Borough President of Manhattan, has been closely involved with growing broadband efforts. Before her election as Borough President, Brewer led the City Council's technology committee, and she testified in hearings about Verizon. Not only has she been active in calling for increased transparency and low-cost options from broadband service providers, but her office also advocated for ensuring that kiosks are placed strategically to help grow adoption rates. "We don't want this to just be an advertising thing for someone to check email," said William Colegrove, Senior Technology Advisor to the Manhattan Borough President. "We want kiosks in zones with digital divide issues, and we're working on raising awareness with community organizations and anchor institutions (W. Colegrove, personal communication, November 9, 2015)." As of this writing, kiosk installation has begun in Manhattan, but the kiosks are not fully functional and will begin as Wi-Fi-only (phone call functionality will come later). The first outer borough kiosks are expected to appear in The Bronx within the first quarter of the 2016 calendar year (Robbins, 2016).

Notwithstanding New York City's experience with Verizon, public-private partnerships, like the city's partnership with CityBridge, are emerging as a solution to the offsetting the high costs of installing fiber. This is particularly the case in low-income communities like the ones served by the ConnectHome program. However, the private sector does not appear to be treating these partnerships as part of their corporate responsibility efforts, but instead views them as a means to revenue generation, mostly through advertising. In the same way that Citibank subsidized the city's bike-sharing program – with the caveat that all bikes be branded with Citibank logos – private sector companies do expect something in return for their efforts to expand broadband access and adoption.

The mutual benefits of such a partnership was ably explained by Andrew Essex, who was the CEO and Vice Chairman of the Droga5 advertising agency through December 2015, when he left to become the CEO of Tribeca Enterprises, which oversees the Tribeca Film Institute and the Tribeca Film Festival. He now describes himself as a "recovering ad executive" but is pleased with the opportunities provided by public-private partnerships, even when it involves the exchange of eyeballs on advertisements for broadband access. "The opportunity is that people don't see ads anymore," he said. "So more money is being diverted to getting people to feel an affinity for something by doing something good. It's better they [advertising agencies] do it through social programs rather than the traditional perspective (A. Essex, personal communication, January 24, 2016)."

Companies not involved in advertising can also benefit from supporting increased broadband access and adoption efforts. Todd O'Boyle is the Deputy Director of Next Century Cities, a nonprofit organization partnering with municipalities throughout the U.S. to deliver faster, reliable Internet access. In explaining how private-sector companies can justify the expense of installing broadband in public housing, he pointed out the benefits a company can reap from the perspective of corporate social responsibility and outreach. "There are many, difficult-to-quantify benefits of broadband," he said. "It's hard to put a dollar figure on how much connectivity helps kids learn, and get ahead in school. We don't know the precise economic impact of making it easier for lower SES (socio-economic status) households to find work, but we do know that increasingly even low-wage, service sector jobs require online job applications. Visionary community leaders understand the benefit of connected communities comes from the public goods they produce – and are creating innovative partnerships to get there (T. O'Boyle, personal communication, January 19, 2016)."

Nonprofit Efforts

However, public-private partnerships are not the only strategy being used in the city to realize complete broadband access and adoption for New York City businesses and residents. Non-profit organizations have also been stepping up to support these efforts, partly with the help of BTOP funds subcontracted to them by the city. One such example is the New York Public Library's hotspot lending initiative, known as the MiFi program, in which 10,000 Sprint NetGear Zing hotspots are loaned out to students, families and individuals enrolled in library programs and adult literacy classes. The initiative, still in pilot mode, has distribution centers in Queens, Manhattan and Brooklyn, in neighborhoods with low broadband penetration rates. Hotspots are loaned for terms of six months to a full year, and while the programs requests that those with home access not borrow devices, they do not require verification of this.

Although hotspots are different from fixed broadband, they are still a low-cost way for people to experience increased digital access at home. Luke Swarthout, Director of Adult Education Services for the New York Public Library, says that the program is an attempt to look at more ways in which libraries can increase digital access by addressing the issue of cost for home service. "Libraries have been involved in broadband adoption for a long time," he said. "The main barriers to adoption are cost of Internet service and devices and a lack of digital skills. Libraries not only provide access to technology, they also offer classes to help patrons understand the role that technology, computers and the Internet can play in their lives. What's new about our project is the effort to address the financial barrier to home adoption. If you walk by a New York Public Library branch after hours it's not uncommon to see someone standing outside with their phone or a laptop, trying to connect via the Wi-Fi bleeding out of the building. Those are patrons who want to be on the Internet, but don't have home service for financial reasons. We have been providing free computer access, Wi-Fi access, classes; this program is about how do we do more." Official data regarding program success have yet to be reported, Swarthout says anecdotal evidence confirms cost, not lack of interest, as the primary barrier for access (L. Swarthout, personal communication, October 20, 2015).

The MiFi program stands apart for its focus on providing hardware to improve home access; the program is not about lending laptops or tablets, which may be of limited use to a household that cannot access the Internet in the first place. The MiFi program does not include skills training – Swarthout says the library is leveraging its existing training workshops – but many other nonprofit organizations focus on expanding access through digital literacy and other trainings.

One such organization is The LAMP (Learning About Multimedia Project), an education organization which brings equipment and hands-on training to low-income communities citywide for the purpose of teaching people to "comprehend, create and challenge" all forms of media. Unlike organizations focusing on specific technology skill sets, such as coding or robotics, The LAMP's focus is on critical thinking skills needed for active media engagement. "Coding is great," said D. C. Vito, Executive Director of The LAMP, "but we think decoding is just as important for meaningful adoption and engagement. Not everyone wants to be the next Steve Jobs, but we all live in a world where technology and media are constantly trying to influence our choices. These basic critical thinking skills can be overlooked – young people often learn coding or programming because those are seen as direct pipelines to high-paying jobs – but they're so vital just for everyday life. A lot of low-income students otherwise don't get this kind of training (D. C. Vito, personal communication, January 27, 2016)."

The LAMP is also a member of the New York City chapter of the Hive Learning Network, which is run by the Mozilla Foundation. The Hive consists of over 70 (mostly nonprofit) organizations offering a variety of digital learning opportunities, related to everything from photography to filmmaking, coding to multimedia art, media literacy to science. The Hive offers networking opportunities for like-minded organizations to meet each other, share best practices and collaborate. Funding is also available through the Hive Digital Media Learning Fund, in which organizations within and outside of the Hive are encouraged to partner to provide programming.

The Hive's strategy for increasing digital adoption and access focuses on providing a network to support organizations dedicated to building skills. "There's still really this need to understand how tech can be useful, how to do basic things, how to create and make things rather than just re-posting, re-broadcasting," said Leah Gilliam, former Director of the Hive NYC Learning Network. "People feel connected and active, but don't always know how to engage and participate online. Whether it's building circuits or building [a] webpage, understanding the infrastructure, history, and utility of a technology can be difficult. There's a lot of work to be done when it comes to empowering people who don't have tech access and/or support to learn and experiment. (L. Gilliam, personal communication, October 26, 2015)."

Mozilla's Hive Learning Network in New York City also collaborates with other city agencies and offices, including the Department of Youth and Community Development, the Department of Education and the Office of the Speaker of the City Council. For example, City Council Speaker Melissa Mark-Viverito's office recently created the Speaker's Committee for Access and Inclusion (SCAI) to encourage New York City Council members to better use technology in serving their constituents and the city at large. The committee is also tasked with raising awareness of the benefits of digital technology across the city as a whole through community input into how these tools and platforms can support their needs. Meghan McDermott, Mozilla's Director for Hive City Strategy, sits on this committee as part of the organization's commitment to efforts that support community engagement in the open web and digital programming.

To demonstrate the importance of creative, hands-on digital programming as part of an inclusion agenda, SCAI and Mozilla agreed to partner on a series of Digital Inclusion events in 2016, which are planned to take place in each borough of the city. The first "learning party" – a hack jam or maker-style event – took place in January 2016 on Staten Island in collaboration with the local Council Member Debbie Rose's Office, the Staten Island Empowerment Zone and several Hive NYC member organizations, including The LAMP. "Our goal in undertaking this collaboration with SCAI is to show what a meaningful strategy for the adoption of digital technology can look like in a community setting," McDermott said. "From this, we hope that council members will be inspired to advocate for healthy funding for digital programming, infrastructure and services. We want them to speak forcefully on behalf of digital inclusion. That would be a win."

The New York City Department of Education

With over one million students, The New York City Department of Education is the largest school system in the United States. As such, its technology needs are great. The federal ConnectEd program, which supports bringing high-speed Internet to public schools, states that school Internet speeds should have fiber optic networks of at least 100 mbps/second, but ensuring these speeds are available to all students at all times is difficult. The Bronx High School of Science made headlines when it asked students not to access the school's Wi-Fi network through their personal phones because the school's network was so strained with non-school related activity. In fact, a report issued by Comptroller Scott Stringer found that 75% of public school buildings have "excruciatingly slow Internet speeds (Harris, 2016)."

One effect of unreliable Internet in schools is that it has left many educators with little incentive to learn how to use it in their classrooms. "If teachers don't think they can count on the Internet, they're not interested in integrating technology," said Lisa Nielsen, Director of Digital Engagement and Professional Learning in Digital Instruction and Information Technology (DIIT) at New York City's Department of Education. She added, "New York City is fortunate to have many tech-savvy teachers. When it comes to the Internet, there are some schools who have what they need, but most people wish they had faster Internet speed."

Part of the setback to increasing broadband access and adoption in New York City schools is the number of students and schools that have so many other unmet, basic needs for learning. Broadband access and adoption can seem like a luxury when a school lacks functional bathrooms, or when a majority of students experience food or housing insecurity. "Once we get them [students] to a place where, like in Maslow's Hierarchy of Need, they're safe, healthy, fed and rested, then the number one learning tool they need is getting them connected to broadband," Nielsen said. "It's the foundation of learning in the 21st century, but if we're not providing basics, we can't move on speed (L. Nielsen, personal communication, January 8, 2016)."

Yet the main reason broadband access is so poor is that, starting in 2011, New York City schools were not been eligible for educational technology funding from the Federal Communications Commission's E-Rate program. Comptroller Scott Stringer estimates the city has missed out on as much as $120 million because of a federal investigation into funds spent by Willard Lanham, a former technology consultant with the city (from 2011 to 2015, the FCC refused to confirm or deny this investigation). In 2012, Lanham was sentenced to 37 months in prison for stealing $1.7 million from the Department of Education. This money was intended to pay for Internet access in public schools, but Lanham allegedly used it to purchase expensive cars for himself and his wife, and to finance construction of luxury homes in Long Island (Harris, 2014).

However, the FCC confirmed in late 2015 that it was investigating whether the New York City Department of Education violated the bidding rules of the E-Rate program, which were intended to ensure that school districts and libraries use price as the main determinant in selecting a service provider. Lanham was named in a report from the Special Commissioner of Investigation for the New York City School District, which found that he gave preferential treatment to two vendors without the DOE's knowledge, and that he stole funds for himself. On December 22, 2015, the FCC reached a $3 million settlement with the DOE, in which both parties agreed to the terms of a Final Compliance Plan. The plan will remain in effect for two years and, among other stipulations, requires the appointment of various compliance officers and establishes an E-Rate Compliance Plan and Training Program in which Carmen Fariña, the DOE Chancellor, must participate (Augustino & Mohr, 2015).

This settlement happened following Mayor de Blasio's announcement in September 2015 of the Computer Science For All (CS4All) program, which will provide computer science to all public school students by 2025. The program will begin rolling out in fifty schools as of the 2016–2017 school year. Training educators for over 1,800 schools to teach computer science will be a significant challenge (Disare, 2016). Bringing reliable broadband services to all of those schools will certainly be another; the city has earmarked $650 million over the next three years to upgrade school technology (Harris, 2016).

One school that may serve as a model for effective technology integration and broadband access is Scholar's Academy, a public school located in Rockaway Park serving roughly 1,300 students from grades 6 to 12 (Agish, 2014). Scholar's Academy shaped its technology mission around the question of how time and place can be removed from teaching and learning. This began with a few teachers using iPod Nanos to record and post lessons online, to every teacher using screencasts to make their direct instruction available anywhere within in the building or at students' homes. "In every lesson or from home, students can pull down the teacher's lesson plan, watch a video or screencast and submit a formative assessment," said Brian O'Connell, Principal of Scholar's Academy. "Wi-Fi in the school is the backbone of this operation (B. O'Connell, personal communication, January 29, 2016)."

By the numbers, O'Connell's operation seems to be effective. In the 2012–2013 school year, 60% of the students were economically disadvantaged, and 62% were in minority ethnic groups. Yet in the same year, 100% of students tested as meeting or exceeding state standards for English, and 97% met or exceeded standards for math. Scholars' Academy students overall performed 37.2% better than other similarly disadvantaged schools in New York State; the school ranked #52 for performance in New York in 2015 (U.S. News, 2015).

O'Connell acknowledged that the school is fortunate to have DITT support for robust Wi-Fi, and that the school's technician is frequently in touch with DITT to manage and monitor usage at the school, particularly during peak timeframes. Still, there are limitations – access points can only handle 25 device connections at one time. If an overload occurs, the devices automatically jump over to the next access point, which can in turn overload that point and create a domino effect. As O'Connell explained, "This is currently problematic in that it prevents us from utilizing large assembly spaces, such as the auditorium, gym or cafeteria, where one may assemble 100 plus students with individualized computing tools to work at their own paces or on individual assignments online. With the looming reality of online testing, these spaces could become great assets to schools as online testing forums."

Lessons Learned and Recommendations

As of this writing, there is still much to be done in order for New York City to reach 100% access to broadband for residents and businesses. The LinkNYC program is in progress, but it's far too soon to understand its true impact or understand how it can scale in other cities. Data is not yet available on the pilot MiFi program, ConnectHome remains in pilot stage, the SCAI is still in the midst of its work to modernize City Council services and the FCC's settlement with the DOE over E-rate funding is barely one month old. The $56 billion proposed takeover of Time Warner Cable by Charter communications is still pending, and the new company would control nearly 30% of the nation's broadband market, in addition to the broadband market for New York City at large (Stenovec, 2015).

Throughout the interviews conducted for this chapter, experts were asked specifically about the impact that policy and legislation have had, or could have, on efforts to increase broadband adoption and access. The responses were varied:

- "The net neutrality decision dictating that all data should be treated the same was pretty important. Edward Snowden's NSA (National Security Administration) revelations also helped to illustrate the range of issues sparked by even the simplest transaction on the Internet. Data privacy, zero-rating, walled gardens, etc. were not super-sexy issues but they became really important, and these incidents pointed to how they impact people's lives and liberties. [We now] Have faces and examples and media to help contour the issues helped spark a wider conversation around access and agency. Suddenly being online becomes intrinsically tied to 'free' services and our agreements with social networks. It became easier to ask, 'What private information are we giving up?'" (Gilliam)

- "I would want to see a policy forcing every fiber and broadband pipe provider to make public fiber availability and its footprint ... There could also be more cooperation from providers. We are also negotiating with private businesses and landlords to run fiber, but more could be done to incentivize right-of-way access ... somehow we have to find a way to deal with more private property disputes." (Colegrove)
- "One of the key things to consider is how much of a digital divide there is in the infrastructure, partly because of how invisible it tends to be, whether that's underground or in the air. Another part is that we've gone backwards. Internet dial-up was based on telephone service, which was universal throughout the city ... Cable is everywhere but not necessarily in the best shape ... the upgrades happening are being done in a highly piecemeal way and decisions are being left to the private market. You now have vast levels of disparity between who is using new fiber and who is using legacy. In addition, the ability for the state or the federal government to tell private companies what Internet speeds they have to serve is under dispute." (Breitbart)
- "If we can't get together to talk about advocating for low-income neighborhoods, how are we going to deal with monopoly? Any time these discussions happen, businesses talk about a chilling effect and usually win with that." (Byrum)
- "There's no law requiring connectivity [in schools], or about how Internet should be delivered. Without that policy, anything a public school district does is linked to available funding sources and to the school district's belief or philosophy about the value of technology." (off-record source within NYC DOE)

Despite some of the legislative barriers described here, New York City should be applauded for its efforts to grow broadband access and adoption, particularly given the unique challenges facing a municipality of its size. Indeed, in some ways the size of the city has enabled it to pursue solutions such as the LinkNYC program and other public-private partnerships; the large companies based in New York are all too happy to leverage opportunities to impact a diverse, densely-populated city that also happens to be a magnet for tourists from all over the world. Whether that impact involves an exchange of personal data or attention to advertising, whether the target is a public housing space or a business district, New York City is a fertile space for experimentation.

But even as these developments in access take place, meaningful adoption depends upon skills-based programming. Just as it may be impossible to come up with a single definition for adoption, concrete definitions and rubrics for what it means to be digitally literate are also difficult to set. However, this does not mean the teaching of these skills can be delayed at the expense of tangible equipment and infrastructural upgrades. Meaningful adoption varies according to one's community, and so it makes sense to invest in community involvement as a path to designing and executing programs that will meet needs. The SCAI initiative is a good example of this effort, but also should be watched closely – the current speaker will complete her term in 2018, leaving the fate of SCAI gains in the hands of her successor, who may or may not share an enthusiasm for digital equity.

CS4All program design and implementation should be watched closely and tightly managed with respect to community needs. The advent of CS4All and funding for educational broadband infrastructure may be an opportunity for forward-thinking schools to finally realize the potential gains of Bring Your Own Technology (BYOT) programming. CS4All curricula should also be watched for how it defines "computer science." Of the two programs announced for the 2016–2017 year, both seem to focus on the mechanics of computing, such as coding, programming and robotics (Disare, 2016). Neither appears to include critical thinking about media and technology communications, or what is commonly referred to as media literacy.

As we at The LAMP have seen since the nearly ten-year ban on cell phones in schools was lifted in 2015, the lack of specific guidance for using personal devices in schools has meant that most schools retain a policy where devices should be neither seen nor heard. The lack of any proactive response to lifting the ban has led to situations such as the one described in the Bronx High School of Science, in which administration is merely reactive.

- Recommendation 1 for New York City Schools: Clearly define and publicize the disciplines included in "computer science," intended learning outcomes and define terms and metrics for achieving basic digital literacy. Consider definitions of the term that are inclusive of multiple interests within the spectrum of computer science (i.e., do not equate computer science only with coding; visual design matters as well). How does current educational policy serve the intended goals of the program, and what needs to change?
- Recommendation 2 for New York City Schools: Select a cohort of elementary, middle and high schools to pilot a BYOT program. Work closely with educators, students and parents to document what serves learning, where professional development help is needed, and how concerns around privacy and security are addressed.

- Recommendation 3 for New York City Schools: Provide support to superintendents and principals in designing plans for the use and management of personal mobile devices in schools, even if a school forgoes the use of implementing a BYOT program. Involve educators, parents and students in determining how personal devices may be used on school property, and appropriate disciplinary action.
- Recommendation 4 for New York City Schools: New York City should lead in passing a law requiring that all public schools meet minimum standards of broadband connectivity as defined by the FCC, with continued technology funding from the School Construction Authority contingent upon meeting those standards.

New York City's nonprofits are vital to the city's goal of achieving 100% broadband access and adoption. Their work, which often takes place at a community level, should be integrated with efforts of local officials and social service agencies seeking to better engage with constituents. But given the varying needs of New York City communities, plans and goals should be developed on a City Council district basis.

- Recommendation 1 for New York City Nonprofits: Individual City Council members should define terms such as "access," "adoption" and "digital literacy" for their districts, in partnership with nonprofit and other community leaders serving their constituents. The definitions should include metrics for 100% success rates, and City Council members should strategize with and support nonprofit community leaders in achieving these goals.
- Recommendation 2 for New York City Nonprofits: District plans should be readily available both in print and online to constituents, together with locally-available resources and trainings for constituents.

The public-private partnerships brokered by the Mayor's Office of Technology and Innovation and DoITT show great promise for the citizens of New York, and may prove to be the most expedient way for vast numbers of New Yorkers, particularly those in low-income communities, to gain access to broadband and the attending opportunities it can provide. However, caution should be taken to ensure transparency as to how individual data are used, and language around expectations needs to be clear.

- Recommendation 1 for the Mayor's Office and DoITT: The City and private-sector companies should take pains to avoid confusion similar to what was experienced in its dealings with Verizon around the term "passed." Strict definitions and success metrics should be stated in every contract, with agreed-upon penalties in place for failing to meet these metrics according to terms.

- Recommendation 2 for the Mayor's Office and DoITT: Prior to embarking on a contract with the city, any private-sector company should be required to publish a draft privacy policy clearly stating how data will be used. Companies should be responsible for holding in-person and/or online question-and-answer sessions open to the public to clarify and address concerns. They should also leave the privacy policies open for a period of public comment before making any final determinations about data usage.

In sum, New Yorkers and others watching its digital initiatives have reasons to be hopeful. Much remains to be seen as to what looks good on paper and what actually works in practice, and unfortunately, we are not strangers to corruption of the sort that contributed to the E-rate scandal. It is clear that the nation's eyes will be on New York City in coming years as we work with innovative solutions and leadership driven to address inequality in all of its forms. We New Yorkers are known for our grit, and we'll need all of it to realize the dream of 100% broadband access and adoption.

Acknowledgments

I wish to thank the following individuals for sharing their time and insight through interviews: Josh Breitbart, Greta Byrum, William Colegrove, Andrew Essex, Hal Friedlander, Leah Gilliam, Meghan McDermott, Lisa Nielsen, Todd O'Boyle, Brian O'Connell, Michael Santorelli, Deb Socia, Luke Swarthout and D. C. Vito.

References

Agish, M. (2014, October 15). Best high schools: Scholars' Academy offers close-knit community, collaboration. *New York Daily News*. Retrieved from http://www.nydailynews.com/new-york/education/best-high-schools-scholars-academy-offers-close-knit-community-article-1.1973897

Augustino, S., & Mohr, R. (2015, December 28). FCC reaches $3 million settlement with New York City Department of Education in E-rate investigation. *Communications Law Monitor*. Retrieved from http://www.commlawmonitor.com/2015/12/articles/universal-service-fund/fcc-reaches-3-million-settlement-with-new-york-city-department-of-education-in-e-rate-investigation/

De Blasio, B., & Roest, A. (2015, June 18). Verizon FiOS implementation: Final audit report. *New York City Information Technology and Communications*. Retrieved from http://www.nyc.gov/html/doitt/downloads/pdf/verizon-audit.pdf

De Blasio, B., & Shorris, A. (2015, April 20). One New York: The plan for a strong and just city. *The City of New York Office of the Mayor*. Retrieved from http://www1.nyc.gov/html/onenyc//index.html

Disare, Monica. (2016, January 29). Kicking off 'Computer Science for All,' city will add AP classes, software programs. *Chalkbeat*. Retrieved from http://ny.chalkbeat.org/2016/01/29/kicking-off-computer-science-for-all-city-will-add-ap-classes-software-programs/

Federal Communications Commission. (2015, February 4). *2015 Broadband progress report and notice of inquiry on immediate action to accelerate deployment* (FCC 15–10). Retrieved from https://apps.fcc.gov/edocs_public/attachmatch/FCC-15-10A1.pdf

Forman, A. (2015, August). NYC's tech profile. *Center for an Urban Future*. Retrieved from https://nycfuture.org/data/info/nycs-tech-profile

Hackett, R. (2015, June 15). States with the most Fortune 500 companies. *Fortune*. Retrieved from http://fortune.com/2015/06/15/states-most-fortune-500-companies/

Harris, E. A. (2014, November 26). New York Schools lose millions amid an inquiry, Comptroller says. *The New York Times*. Retrieved from http://www.nytimes.com/2014/11/27/nyregion/new-york-schools-missing-out-on-millions-in-technology-funding-comptroller-says.html?_r=0

Harris, E. A. (2016, January 13). Bronx science bans cellphones from Wi-Fi as students devour it. *The New York Times*. Retrieved from http://nyti.ms/117UDgk

Horrigan, J., & Duggan, M. (2015, December 21). Home Broadband 2015. *Pew Research Center*. Retrieved from http://www.pewinternet.org/2015/12/21/home-broadband-2015/

New York City Department of Parks and Recreation. (2013, July 17). More public Wi-Fi is launched At 32 parks across the five boroughs. *The Daily Plant, XXVIII* (5902). Retrieved from http://www.nycgovparks.org/news/daily-plant?id=22949

Office of the Mayor. (2015, July 16). Mayor de Blasio announces up to $10 million investment in free broadband service for five NYCHA developments. Retrieved from http://www1.nyc.gov/office-of-the-mayor/news/491-15/mayor-de-blasio-up-10-million-investment-free-broadband-service-five-nycha#/0

Robbins, C. (2016, January 5). Brace for the "fastest Internet you've ever used" at these free sidewalk kiosks. *Gothamist.com*. Retrieved from http://gothamist.com/2016/01/05/linknyc_Wi-Fi_2_fast_2_furious.php

Silbey, M. (2015, June 29). Who's feeding fiber to LinkNYC Hotspots? *LightReading*. Retrieved from http://www.lightreading.com/gigabit/gigabit-cities/whos-feeding-fiber-to-linknyc-hotspots/d/d-id/716641

Stenovec, T. (2015, May 26). Here's what the $56 billion charter and time Warner cable merger is really about. *Business Insider*. Retrieved from http://www.businessinsider.com/charter-time-warner-cable-merger-is-about-broadband-2015-5

United States Department of Commerce National Telecommunications & Information Administration. (2015, December). Broadband USA grants awarded map: New York. Retrieved from http://www2.ntia.doc.gov/New-York

U.S. News. (2015). Scholars' academy overview. Retrieved from http://www.usnews.com/education/best-high-schools/new-york/districts/new-york-city-public-schools/scholars-academy-91638

Verizon. (2015, June 12). Response of Verizon New York Inc. to the draft audit report of the New York City Department of Information Technology and Telecommunications. Retrieved from http://legistar.council.nyc.gov/LegislationDetail.aspx?ID=2474216&GUID=BA4CB817-1D1E-4BFB-B954-985E5BCB280B&Options=&Search=

Verizon. (2015, October 14). New York City Council Hearing – Verizon Testimony Submissions. Retrieved from http://legistar.council.nyc.gov/LegislationDetail.aspx?ID=2474216&GUID=BA4CB817-1D1E-4BFB-B954-985E5BCB280B&Options=&Search=

Wimer, C., Garfinkel, I., Gelblum, M., Lasala, N., Phillips, S., Si, Y., … Waldfogel, J. (2014). Poverty tracker – monitoring poverty and well-being in NYC. *Report 1. Columbia Population Research Center and Robin Hood.* Retrieved from https://www.robinhood.org/sites/default/files/user-uploaded-images/RH_PovertyTracker_121214.pdf

Zara, C. (November 24, 2014). NYC free Wi-Fi plan isn't really free: Data collection, advertising space at the heart of CityBridge. *International Business Times (U.S. Edition).* Retrieved from http://www.ibtimes.com/nyc-free-wi-fi-plan-isnt-really-free-data-collection-advertising-space-heart-1727899

CHAPTER TWELVE

Regulating the Place of Food Trucks

GARY GUMPERT
Urban Communication Foundation

SUSAN J. DRUCKER
Hofstra University

It is hardly necessary to discuss the clear link between food and memory. Proust's "petites madeleines" demonstrate the strong impact of food on memory. The taste and aroma of the street vendors' roasted chestnuts during New York's holiday season punctuate the memories of many. The tale of street vendors, particularly those selling food, is one rooted on urban lore and law. These businesses serve as sites of contact and interaction between the vendors and the public.

The Daffodil Massacre as told by Jean Merrill in The Pushcart War—a story told about a conflict occurring between the small pushcart vendors as the large trucks crowded them out in the narrow east side streets of New York. It was a children's tale originally written by Jean F. Merrill in 1964 but about future events in 1976. Because we had to replace the original torn and tattered edition, we recently ordered a new issue published by *New York Reviews Children Edition* and discovered that in the new edition Jean Merrill's introduction was dated October 14, 2016—although she had died in 2012. We looked for an explanation and it came in a fascinating *New Yorker* column by Peter C. Baker titled "The Enduring Appeal of 'The Pushcart War'" (September 17, 2014). Baker explains "In the Books first edition, the story unfolds in 1976. When it was republished in 1974, the setting was changed in 1986." The latest edition published in 2014 has two forwards, the first by Professor Lyman Cumberly of New York University—actually a fictional persona used as part of the stories' journalistic style.

There is always a hidden reality about our memories and a fictionalized landscape. Peter Baker explains:

> For Merrill, the book-which takes the form of piece of narrative journalism published ten years after the war ended—was something of a revenge fantasy. Living in Greenwich village during its mid-century period of intense development, she came to hate all the loud, smelly, hulking trucks off the roads. And she fantasized about deflating their tires. Things seemed likely to get worse: Robert Moses, the city's "master builder," had plans (never realized) to construct a ten-lane elevated highway that would cut through downtown. Later, Merrill traced the origin of "the Pushcart War" to her realization that what she felt about the trucks "is what everybody feels about bullies."

Pushcarts and Chuck wagons are the precursors of modern mobile street food vendors. Chuck wagons followed trail-driving cowboys in the Wild West after the Civil War. The urban pushcart served urban dwellers and workers in cities like New York and Chicago. "Garment workers, construction men and delivery boys relied upon the carts for cheap and filling nourishment in the middle of busy days" (Butler, 2014). Pushcarts and street vending had long been a reality in New York, particular in the lower East Side at the turn of the century. It was a world of immigrants seeking to survive in the tenements and peddling was a convenient option, but they were not always welcomed by all.

Street vendors are established fixtures in many neighborhoods, part of the fabric of many communities. They are small businesspeople often struggling to make ends meet. Most are immigrants and people of color; diverse people selling diverse foods. Mobile vendors are embedded in the culinary scene of cities. They are even reviewed and rated by food critics.

Whether serving ethic foods per se, or interpersonal contacts, street food offers opportunities for interpersonal and intercultural communication. Street vendors teach ethnicity through consumption. M. F. K. Fisher said "With our gastronomical growth will come, inevitably, knowledge and perception of a hundred other things, but mainly ourselves" (1988). Casey Man Kong Lum and Marc de Ferriere le Vayer have asserted that "foodways is an important material and symbolic marker of identity, race and ethnicity, gender, class, ideology and social relations" (Lum and De Ferriere le Vayer, 2016). Preparing, consuming, and sharing food is part of daily life's communicative acts so important the way we interact with others.

Quotidian acts of buying milk, ordering morning coffee, standing in line at the bakery, purchasing a taco from a food truck are all significant to urban communication. Food is a symbol in our creation of meaning. Greene and Cramer conceptualized food as a "a nonverbal way we create, share and transform meaning with others" (Cramer and Greene, 2011). What we eat, where we eat and with whom we eat create and sustain relationships between individuals, and attachments to communities, and even countries. The sites in which food is obtained and consumed offer opportunities for interaction. Whether at the dinner table ordering through a window, communication food brings people together.

The growth of mobile food vending has garnered attention as a means "to expand economic opportunity, and enrich communities by providing a diverse array of food options that augment existing brick and mortar restaurants" (Food on Wheels, 2013). They are viewed in terms of policies and regulations associated with "economic activity, public space, public health and public safety" (Food on Wheels, 2013) but never their varied communicative functions, the subject of this essay.

The general path of street selling—from pushcart vendors to food carts to food trucks—is the tale about the changing nature of the street as it became transformed into a thoroughfare occupied primarily by motor vehicles. In that sense regulation primed by the automobile is a necessary tale of regulation required for traffic flow and control. The transmogrification of street food vendors involves the convergence of several elements: the development of the automobile accompanied by the changing design and dominance of street and traffic; an urban work force requiring by its necessary mid-day food consumption; and the need to regulate vendors and their products for economic, traffic and health reasons. This complicated process of growth and change of food culture is connected by the interface of that industry with a growing and diverse immigrant population who sought some means of making a living and initially the vending of food provided such a means. Then and now the pushcart and the food truck provided a quick and different cuisine for a transitory population. It has been a novel and perhaps somewhat exotic food experience—for many it became initial exposure to the alien.

Mobile Street Vending and the Food Phenomenon

There has always been a link between urbanity, food, immigration and regulation. While examining urban foodways could certainly encompass street food in general, the unique social, media and regulatory issues associated with mobile street food vending warrant the focus of the current examination. As of 2015 there were an estimated 4,130 food trucks in the United States worth $1.2 billion (Field, 2016). Social media reviews and the ability to broadcast or track location has been associated with the more recent rise in food truck popularity. In many cities websites offer the mean of tracking location of a favorite truck or cart. One can turn to http://mobile-cuisine.com/find-food-trucks/ to locate a favorite truck. Hunting for that truck—there's an app for that which is "Roaming Hunger Food Truck Finder" available at the iTunes App Store. Zagat (restaurant raters) offers an Interactive Food Truck Finder. Available for Android or Apple device is

Figure 12.1. Street vendor in lower Manhattan.

Mobile Nom Food Truck Finder at https://www.mobilenom.com or Food Truck Finder USA at http://foodtruckfinderusa.com/. City specific searches are aided by sites such as ChicagoFoodTruckFinder.com. Hungry in New York City go to Roming Hunger at http://roaminghunger.com/food-trucks/ny/new-york/ or NYCTruckFood.com. Bloggers provide reviews and location data (e.g. What the Trucks https://whatthetrucks.com/blog/tag/find-food-carts-food-truck-finder/).

Food truck urbanism has taken off. Currently, food trucks and carts are found in cities across the United States and well beyond. Cities known for their food trucks include Chicago, Los Angeles, Miami, Houston, San Francisco, Boston, Portland, Oregon and Washington D.C where even the President has his personal favorite (i.e. D.C. Empanadas) (Fabian, 2011). owadays, food trucks are even used to feed people at special events, like birthday parties, weddings, school dances, and public gatherings. Another new trend is food truck festivals or events at which several food trucks cluster creating further unique environments of communication.

Food trucks and carts are popular in many cities around the world from Paris, Geneva, Helsinki and Berlin to Hanoi, Montreal, London, Cape Town (Field, 2016). They frequently furnish diverse cuisine from around the world. Food trucks have even ushered American cuisine into the culinary capital of the world: Paris. According to Paris Mayor Anne Hidalgo, "Food trucks are a new culinary trend and must get real recognition as part of French gastronomy" (Berton, 2015).

Paris guidebooks offer locations, menus, reviews and contact information (Bedeau, 2015). Bud's Deli offers classic New York deli featuring pastrami.Burger and bagel trucks offer a taste of New York where, according to Julia Moskin of the New York Times, "Among young Parisians, there is currently no greater praise for cuisine than 'très Brooklyn,' a term that signifies a particularly cool combination of informality, creativity and quality" (Moskin, 2012). Burgers are featured by the popular Le Camion Qui Fume (The Smoking Truck), owned by California native Kristin Frederick. Cantine California offers burgers and tacos stuffed with organic meat. Burgers are also featured at Wagy Burger Bus, On Mobile Burger, Hipster Food Truck, Streat Parade and others. Bagels are the thing at Bügelski, a food truck inspired by the delicatessen. In the mood for BBQ, BBQ Corner. American food is not the only exotic cuisine offered in Paris with alternatives including Une Currywurst Paris-Berlin, Mexican food at Senor Boca, African food can be found at Black Spoon. Italian food is featured at Mozza & Co., Tommasino Pizza, and After Eleven while Asian food is the draw at Tooq Tooq, Thai la Route, Le Camion Bol (Bedeau, 2015). Other trucks offer fish and chips, Vietnamese bánh mì and Venezuelan arepas (Berton, 2015).

A further sign of the acceptance of this trend came when Paris gave permission to a limited number food trucks allowing them to regularly operate within 40 designated areas of the city for a one year period (Berton, 2015). "56 trucks were chosen from 158 applications, based on food quality, respect for the environment, looks, plus the potential to create jobs" (Berton, 2015). Food trucks had been permitted outside Paris at resorts and beaches. As with other cities, food trucks became popular in areas such as the business district of La Défense, frequented by office workers who either lacked other eating options or sought to forego a sit down restaurant during the workday. Prior to this 2015 edit, the process to get authorization to park food trucks within was almost impossible. "The main reason for the reluctance, apart from gourmet snobbery, has been the fear that with much lower costs of production, those culinary vehicles pose unfair competition to 'sedentary' restaurants" (Rodriguez, 2015). Authorities plan to review the scope of the permissions after one year (Berton, 2015).

The tale of Paris suggests the introduction of culinary tastes and new social spaces offering intercultural interaction through food and proprietors. One is reminded of the perpetual accompanying regulatory issues as well.

The evolution of the quick food experience begins with the individual small immigrant but over a period of 100 years evolved into the food truck industry. There is a long and complicated food cart vending history, but the contemporary food truck emerged in Los Angeles. In the 1970s and 1980s mobile vending in Los Angeles was primarily a business for recent immigrants especially run by Mexican immigrant (Frommer & Gall, 2012). The regulatory environment was not supportive of these vendors and there was an effort by some city officials to shut down mobile vending. Some of these early vendors worked with civil-rights lawyers to fight the regulations one of which prohibited food vending within 100 feet of a restaurant's front door (People v. Ala Carte Catering Co, 1979). In 1984, "the California Legislature passed a landmark provision telling cities that they may only regulate mobile vending 'for the public safety.'" (Frommer & Gall, 2012, p. 12). Subsequently other legislative measures prohibited banning mobile vending for any reason (Frommer & Gall, p. 12).

Within this background by the 1990s the Los Angeles area was an important place for mobile vending as evidenced by the popularity of the Kogi truck, a catering van that served a fusion of Korean food and Hispanic food (short-rib tacos, kimchi dogs). it was to some degree a successor of the mobile fleet of taqueros (taco sellers) and loncheras (traditional taco trucks) prevalent in the L. A. area.

The Kogi truck was the result of the partnership of Roy Choi and Mark Manguera in 2008. Roy Choi was honored in 2010 by *Food and Wine Magazine* as one of the top ten "Best New Chefs"—unusual for a food truck chef to be honored int his way. In 2012 Jonathan Gold, in Smithsonian Magazine, noted that "Kogi represents mobility in a city that worships mobility; it is a vehicle for traversing lines of race, class and ethnicity; it is selling a social experience as much as it is selling Blue Moon mulitas and Blackjack Quesadillas" (Gold, 2012).

Today mobile food vendors, whether pushcart of motor vehicle, must be licensed and authorized with their equipment and size specified and governed one or more jurisdictions, by season, by type of vending, and special areas for vending.

Regulation of Food Carts and Trucks

The longtime tradition of mobile food vendors in the urban settings has always brought with it the issue of appropriate regulation. In the United States that history has been dated back to 1691 when in New Amsterdam (New York City) street vendors selling food from push carts were regulated (Mobile Cuisine, http://mobile-cuisine.com/business/history-of-american-food-trucks/).

Urban street vending is encouraged in some places and restricted in others. In recent years, street vending has been an explosive issue as municipal officers; health inspectors and neighborhood civic associations react differently to this use of public space. "Many municipalities have updated vending legislation in the past decade with the boom of mobile food vendors and the emergence of a food truck culture" (Ngo, 2012, p. 17). Street food vendors battle brick and mortar establishments who view vendors as unwelcomed competitors and interlopers. Some vendor advocates argue many laws are the result of lobbying by brick-and-mortar restaurants "which argue that since food trucks don't have the same costs in terms of rent and property taxes, they amount to 'unfair competition.'" (Frommer & Gall, 2012).

Some cities limit carts and trucks from operating within 200 feet of permanent outlets serving food and drink. Numerous (and sometimes conflicting) regulations are required by the municipal departments including Health, Sanitation, Transportation, Building and Consumer Affairs. Refuse requirements require that food trucks clean up trash making vendors responsible for keeping specified areas around areas of operation clear of litter. Besides physical restrictions and use restrictions embodied in zoning laws, regulations include caps on the number of licenses available in any one city, fees for licenses and violations and rules on hours of operation. Enforcement can be erratic. (Basinski, August 26, personal communication; Frommer & Gall, 2012).

There is a long legal tradition in regard to the availability of the street as a public site for selling. Does the public have an inherent right to the street or to the sidewalk? What is public? What is private? The answer to such questions is simply not clear.

Regulations governing this environment are of two types: a) primary regulation: regulation primarily aimed at communicative activity and b) secondary regulation: regulations aimed at non-communicative activity but which affect communication. Primary regulation of food trucks and carts can be associated with the signage of the vehicles themselves or the communication taking place in and nearby. However, the indirect regulations are most significant when considering the regulation of food trucks and carts within a U.S. context. The indirect regulations include land use regulations, primarily, zoning laws which address take-out and delivery service, and vending machines, plus the segregation of functions or design districts which all influence communication patterns. Other indirect regulations address loitering, public nuisance including noise, the licensing of alcoholic beverages, minimum drinking age and driving age, the regulation of smoking in a public place.

The bureaucracy of vending regulation is complex and specific. At the present time, the problems of regulating food trucks and carts have become an increasing litigious issue wrought with intricate and bureaucratic tangle of regulations. In the United States setting aside the necessary health rules and regulations that govern the selling of food, the vendor of books and art are inherently protected by the First Amendment. Some vendors have been dubbed "First Amendment" vendors because they sell constitutionally protected printed materials (e.g. newspapers, books, magazines, dvds, etc) or art are legally allowed to set up anywhere that any type of sidewalk vending is permitted without a license. The U.S. Supreme Court has held that "a law subjecting the exercise of First Amendment freedoms to the prior restraint of a license, without narrow, objective, and definite standards to guide the licensing authority, is unconstitutional." (Shuttlesworth v Birmingham, 1969]). Therefore requiring First Amendment vendors to obtain a license would impermissibly restrict the exercise of their First Amendment rights. Reasonable time, place and manner restrictions of First Amendment activity are permitted and can be applied to vending on a public street. First Amendment vendors must comply with city restrictions and tax laws. This This essay makes the assumption of a connection between food and communication. We see it as a given. But despite the communicative properties of food and food consumption, food vendors do not enjoy the added protections provided First Amendment vendors.

Location, Location, Location

Location restrictions take several forms. Many cities have restrictions on where food trucks may stay and may require that they remain a certain distance away from their brick-and-mortar competitors. Some cities zoned out portions of the city prohibiting vending (Norman, Frommer, Gall, & Knepper, 2011). Distance restrictions may require food trucks to stay 100 feet from schools (e.g. Los Angles) (Frommer & Gall, 2012) or at least 20 feet from a building's entrance or exit (e.g. New York City) (What Mobile Food Vendors Should Know, no date). Some cities restrict the duration a vendor may stay in one location making the mobility of vending a key either limiting the specific amount of time they can remain or requiring they keep moving without returning to any one location within a certain amount of time (Frommer & Gall, 2012). The National League of Cities explored food carts and trucks in "Food on Wheels: Mobile Vending Goes Mainstream," a study of thirteen U.S. cities conducted in 2012. The study revealed moderate or less restrictive parking durations to be the pattern in cities with five cities having no time limits and three with a 45 minute or less limit. But, they noted it is cities with more restrictive limits that often have lax enforcement of these regulations (Food on Wheels, 2013).

As the Institute of Justice Report "Food Truck Freedom" noted:

> Forcing vendors to move regularly makes it difficult, if not impossible, to run a profitable business. Short time limits also pose a safety hazard, since it pressures cooking trucks into moving before their equipment has completely cooled. And by requiring trucks to constantly be on the road, laws like these make congestion worse, not better. (Frommer & Gall, p. 12)

Duration restrictions pose a unique challenge for some types of food trucks or carts where cooking takes place. The Institute of Justice conducted a survey which exposed an additional challenge for operators who require time to set up, cook and close the kitchen in order to move. "Trucks also complained about the harm to their business's reputation when they have to turn away customers who have patiently waited in line" (Frommer & Gall, p. 19).

Unlike food carts parking on the street is a problem for food trucks. Some cities prohibit food trucks from parking in metered spots (e.g. Pittsburgh). "Most cities in the United States allow food trucks to pay for and operate from metered parking spaces for the amount of time listed on the meter" (Frommer & Gall, p. 12). However, in New York City the New York City Department of Transportation Regulations §4–08(h)(8) states that "[n]o peddler, vendor, hawker or huckster shall park a vehicle at a metered parking space for purposes of displaying, selling, storing or offering merchandise for sale from the vehicle." A food truck sued, arguing that its food was not "merchandise" for purposes of the law but the litigation ended with a finding in favor of the city regulation (Monroy v. City of New York, 2012).

Pedestrian traffic flow in crowded areas can be the basis of regulation. In New York City food trucks are kept out of the busy theater district of midtown (Rules and Regulations for mobile food vending, 2016) whereas in other cities limitations are specifically placed on vending in areas which intensify congestion or create a hazard. So under the Las Vegas City Code §6.55.070(A)(2) no mobile food vendor shall "[v]end in a congested area where the operation will impede pedestrian or vehicle traffic" (Frommer & Gall, 2012).

On the Streets of New York

In New York City, one of "the first law(s) regulating food carts was known as the Thirty Minute Law, wherein a pushcart had to re-locate every thirty minutes. That was a difficult law to enforce (especially in the overcrowded poorer neighborhood of the Lower East Side) and the law was probably disobeyed more than it was obeyed. In fact, the very first pushcart market was established on Hester Street in 1886 when four Jewish peddlers decided to stay put for much longer than 30 minutes. These markets grew in popularity across lower Manhattan. While many organizations made attempts, there was little regulation at these markets and corruption, uncleanliness, and chaos ran rampant" (Hoffman, 2011).

The regulatory web of mobile food vending in New York City includes Article 89 of the New York City Health Code, Chapter 6 of Title 24 of the Rules of the City of New York and Subchapter 2 of Chapter 3 Title 17 of the New York

City Administrative Code. In New York, it's the Department of Health (for food vendors) and Department of Consumer Affairs (for non-food vendors). But the renaissance of food trucks has met with more recent hurdles and their popularity has slowed down because of myriad obstacles placed in their way. The number of trucks has remained fairly static while the total of food carts has increased, partly due to the technical modernization of the carts and their obvious greater mobility. The food truck business is a difficult one.

> ... The business stinks. There are numerous (and sometimes conflicting) regulations required by the departments of Health, Sanitation, Transportation and Consumer Affairs. These rules are enforced, with varying consistency, by the New York Police Department. As a result, according to City Councilman Dan Garodnick, it's nearly impossible (even if you fill out the right paperwork) to operate a truck without breaking some law. Trucks can't sell food if they're parked in a metered space ... or if they're within 200 feet of a school ... or within 500 feet of a public market ... and so on. (Davidson, 2013)

In a 2013 Article, Adam Davidson points out:

> The food-truck business, I realized, is a classic case of bureaucratic inertia. The city has a right to weigh the interests of food-market owners (who don't want food trucks blocking their windows) and diners (who deserve to know that their street meat is edible, and harmless). But many of the rules governing location were written decades ago. In the '80s, the city officially capped the number of carts and trucks at 3,000 (plus 1,000 more from April to October). (Davidson, 2013)

However, it was estimated that by 2015 there were 8000 mobile food vendors in New York City (a combination of official and illegal operators). The term "mobile food vendors" includes any food vending unit that does not have a permanent foundation.

This cap has been the subject of a long fought battle. In October 2016 the New York City Council Committee on Consumer Affairs took up the Street Vending Modernization Act which would double the number of food vendor permits over seven years with 5% of new permits being set aside for veterans and the disabled. The Act was sponsored by Melissa Mark-Viverito, the Council speaker, and at least six other members. The aim is to address the black market for permits and improve enforcement of regulations, especially in congested areas (CUNY).

The Act would create a new dedicated vendor law enforcement unit to make sure vending rules and regulations are followed and applied consistently. The Act also focuses on initial enforcement efforts on congested areas and around supermarkets. It seeks to "improve compliance by installing street signs on blocks without legal vending, creating an app with a satellite view of legal vending spots, and requiring vendor training" (New York City Council Hearing, 2016).The proposed act would also establish a street vendor advisory panel to monitor enforcement and new permit rollouts, introduce a pilot program to create model vending zones and review outdated rules governing food vending (New York City Council Hearing, 2016). Preference would be given to vendors on the city's waiting lists for permits; approximately 2,500 people are currently on the list for full-time permits.

Technically, a permit for a food cart or truck is not transferable, but Andrew Rigie, executive director of the N.Y.C. Hospitality Alliance, said that vendors regularly pay permit holders something like $15,000 to $20,000 to lease their certificates for two years. Legally, the permit holder becomes a junior partner in the new business (Davidson, 2013).

Other regulatory considerations fall under the New York City Health Code which NYC health code applies equally to restaurants, catering as well as mobile food vending establishments. According to Daniel Kass, formerly of the NYC department of health who was involved with enforcement issues, many of the regulations are designed to protect the public from food contamination. The Health Code does stipulate the kind of equipment required depending on the on type of food processing going on or if a cart or truck. For example, "Donut, coffee, hot dogs, bagel carts most basic … these require minimal handwashing. As soon as there is cooking or food stored under refrigeration the number of applicable regulations apply" (Kass, November 4, 2016, personal communication).

There has been a technologizing of food trucks and carts. Among the technologies is signage which has become brighter and blinking and more garish. Codes are silent as to lighting and noise or music emanating from carts and trucks, but many argue it should be treated as visual blight. There are noise codes which apply to the area of the truck or cart, but nothing to limit food vendors from using bright digital advertising (Kass, 2016).

In addition to the regulatory maze in which the food truck entrepreneur find themselves New York City streets must be maneuvered in order to find an advantageous site from which to operate. "We leave at 4:30 a.m. to get a spot on the street, but now I can't get to the space where I've been parking for the past 18 months because others are leaving at 3:30 a.m." (Fickenscher, 2014). New York City has no blanket restriction on hours of operation, instead restricting vending during certain hours only at specified locations, unlike some cities with very specific limitations on hours of operation (e.g. Phoenix prohibits operation after dark (Frommer & Gall, 2012).

Not all mobile food vendors are created equally. There is a difference between a food truck and a food cart. Carts are on sidewalks while trucks are driveable and parked curbside. Whether a cart or truck, the space around these conveyances are communicative spaces and sites of interaction. Both require the same permit and both are inspected by the same agencies but there are few differences between the two. Food trucks permits require a larger water tank. Since trucks must be parked curbside, parking is a significant challenge. When vendors cannot stay in one regular location they develop a clientele which can be lost if they are required to move from their regular location.

Vending and the Privatization of Public Space

Much has been written with regard to the privatization of public space. In writing about private public spaces urban planners Jeremy Nemeth and Stephen Schmidt suggest considering "publicness' along three distinct but interrelated dimensions: ownership, management, and uses and users" (Nemeth & Schmidt, 2011, p. 5).

The dimension of ownership is usually answered by whether the space is owned by a government body (public) or a private individual or corporation (private). From this dimension we would assume we could learn about the management and operation of the public space. "Typically, ownership is directly related to operation: publicly owned spaces are usually publicly operated; privately owned spaces are normally privately operated" (Nemeth & Schmidt, 2011). But this may not always be the case.

Privately owned public space (POPS) became a popular term often associated with the work of Harvard Professor Jerold S. Kayden in his book *Privately Owned Public Space: The New York City Experience* (Kayden, 2000). POPS are spaces that, although privately owned, are legally required to be open to the public. This form of privatization of public space created public spaces through public-private partnerships. POPS were seen as an alternative to publicly owned spaces in meeting the public need. Beginning in 1961 in New York City zoning variances were granted to allow hundreds of office and residential towers to built beyond the limits of zoning law floor area ratios. In exchange for these valuable variances or bonus incentives, private owners were required to provide public spaces. These public spaces have taken a variety of forms from outdoor plazas and indoor atriums to arcades legally required to be open and accessible to the public. This trade-off has resulted in approximately 20 million square feet of additional retail/residential spaces for 80 plus acres of POPS concentrated in Manhattan's midtown and downtown with a fair number in East Midtown and the Upper East Side. Kayden profiled 503 POPS in Manhattan revealing an unevenness in the quality and maintenance of the resulting public spaces.

The resulting spaces have had mixed results. "At their best, these spaces marry aesthetics with function, offering unique physical and social environments within a densely packed urban center. At their worst, they are barren, unusable surfaces or privatized-by-management spaces that diminish the spirit underlying the laws that created them" (Kayden, 2000). Among the most famous of these POPS is Zuccotti Park in Lower Manhattan, where a protest camp was established as a key component of the Occupy Wall Street protest. Some of these POPS are designed to appear to be open to the public while others give the impression that they are part of a privately owned building open only to residents. Some have gates delineating the space while others have been created to be uninviting offering few amenities like seating.

POPS allow a municipality an alternative way of providing public open space while shifting the direct costs associated with owning and maintaining these spaces. Typically rules of conduct signs are posted to whose enumerate prohibited activities. Critics have noted these rules "demonstrate the ongoing challenges of balancing competing interests in the private provision of public goods" (Woodward, 2012). The private owner has a great deal of latitude in establishing rules including who will be given permission to vend in this type of "public space." As an increasing percentage of public space became privately owned, vendors sought access to these spaces. This type of Private use of Public Space (PUPS) could in large part be determined by private owners.

Land ownership, has a significant impact on who gets permission to sell and what type of vending can take place. As private ownership of public spaces become increasingly important in a city like New York, the power of the private owner in determining access and conditions has become an issue for vendors.

A comparatively new variation on a theme of privatization is reflected in the increased private power over publicly owned public spaces. New plazas in the city have been created that are PUBLICLY OWNED but administered in a manner that is more privatized than in the past (e.g Times Square Plaza). This important trend in public space began under the Bloomberg administration and has continued under de Blasio administration. Among the dimensions of publicness noted earlier was management. "This dimension refers to the manner in which a space is controlled and maintained, and specifically refers to the methods by which owners indicate acceptable uses, users, and behaviors. Management techniques range from including features that encourage freedom of use, access, and behavior (such as making seating available) to providing elements that discourage use and control access and behavior, such as the presence of panning surveillance cameras or armed security guards. Understanding the various approaches to spatial management is important because behavioral control often has broader consequences related to the degree of inclusiveness and social diversity of a space" (Nemeth & Schmidt, 2011, p. 11). According to Franck and Paxson (1989, p. 133), "who controls a public space, how they do so, and how they attempt to make the space safe and secure" are important components of this management and control dimension" (Nemeth & Schmidt, 2011).

In New York City this important trend began under the Bloomberg administration and has continued under the de Blasio administration. In New York City *Business Improvement Districts* (BIDs) are created by the City Council. Local representatives often promote the establishment of BIDS in their districts. All services are funded by a special assessment paid by property owners within the district. The City's BIDs invest over $130 million annually in programs and services for neighborhoods across the five boroughs (Business Improvement Districts, 2017). The result has been that BIDS are becoming an ever more potent power. New York City BIDs are chosen to administer an area or plaza and in exchange the BID is bestowed with certain management rights. (e.g. to say how a place is used or which vendors may operate there).

In an interview with a representative of the The Times Square Alliance the supplementary nature of services was emphasized. When asked about policy on such matters as street vendors, street performers and signage (all highly regulated features of the urban landscape), the power of the city was the fallback position. When interviewing representatives of those regulated such as food vendors, the power and discretion of the BIDS is underscored.

Critics assert that some city governments are relinquishing their responsibilities and entrusting private entities with the job of not only financially supporting but maintaining and even managing urban public spaces such as parks. In New York City there has been alarm voiced regarding this new so called "creative" model for public spaces. Demands have been made for a different model, "one that reflects a democratic, fair, transparent and inclusive process" (Swan, 2013). Street vendors feel the managerial power of BIDS. In New York, Sean Basinski of the Street Vendors Project notes:

> All over the city the issue they is that they don't just set aside space; a BID is chosen to administer an area or plaza and in exchange the BID gets rights. (e.g. to say how to treat vendors or which vendors are allowed); what is happening is that the city takes a street and creates a plaza; they call it a plaza so one would think the vendors should have more rights there except what is happening is that BIDS don't want vendors who are not paying any rent to the BID (eg. 23rd and Broadway by Madison Square park) with kiosks paying rent to the BID that controls it and takes rent from kiosks; those are run by companies NOT immigrants and small business vendors. (Basinski, August 26, personal communication)

There are 72 Business Improvement Districts in New York City created by the City Counsel meaning the BIDS are becoming an ever more potent power in determining which vendors get to set up a kiosk or cart on public property. Some BIDs have published their own guidelines for vending (e.g. SOHO Broadway) (SoHo Vending Guidebook, 2016).

The BIDs have become involved in the Street Vendors Project campaign to lift the cap on the number of vending licenses. The New York City Business Improvement District Association issued a statement saying "We oppose any proposal to lift the cap on vending permits in the absence of a comprehensive, deliberate, and research-based study, with input from public stakeholders." (New York City's street vendors work to decriminalize sidewalk selling, 2015).

Cart vs. Truck

Carts are on sidewalks while trucks are driveable /considered curbside.

Both require the same permit both have inspectors and there are few differences but a truck has a bigger water tank; so small permit difference but exist but not too many. For trucks the main problem is that there is no place to park legally. The 1965 traffic ordinance says no hawkers, hucksters, etc; can use a metered spot and nearly all spots are metered. However the were not being enforced. According to Sean Basinski, the Street Vendor Project assisted in bringing a lawsuit on the parking issues claiming a right to work anywhere, however they lost the case (Basinski, August 26, personal communication). Tickets are still given or vendors are told to move. It is extremely precarious to have a food truck in NYC. Vendors create a clientele and then told to move so you lose the clients. Some streets prohibit street vendors completely in an effort to keep traffic moving (Rules and Regulations for mobile food vending, 2016).

In 2011 a proposed New York City law attempted to "overhaul and streamline" the street vending regulations The revised law "would purportedly have opened almost all city streets and avenues to street vendors, subject to a statutory scheme involving licensing, allocation of space for vending, and a priority system for vendors seeking to vend at any given location." The proposed met resistance and the New York Civil Liberties Union testified opposing the change, because the modification would impose a licensing system antithetical to the First Amendment. The NYCLU opposed the revision because it would "unduly burden protected First Amendment activity." Inherent is the distinction made between those vendors who sell or distribute "newspapers, periodicals, books, pamphlets or other similar written material." (The Regulation of Street Vending, 2011).

Implicit, however, is the recognition for the need of a licensing and regulatory system that would govern those vendors not covered by the First Amendment especially those selling food. The distinction while fairly clear are clouded by a further degree of ambiguity between private and public space and this is a particularly murky area. In a recent meeting with Sean Basinski, Director of the Street Vendor Project (part of the Urban Justice Center) the matter of the control of public space was raised as an important issue.

"The most unusual site of operation is these days, the aptly named Green Pirate Juice truck spends four days a week on the 12th floor of the industrial style Starrett Lehigh Building at 601 W. 26th St.—yes, it drives right into the service elevator—where it serves freshly squeezed juices to the mostly fashion designer tenants. The rest of the time, it's hired for events by Silvercup Studios, concert promoters and others" (Basinski, August 26, personal communication).

While some of the food truck entrepreneurs have given up, other have moved culinary delights to older style restaurants. The idiosyncratic unique cultural domain is being reshaped by "the invasion of the big boys." "From around 2010 to 2012, restaurant franchises and fast food chains began co-opting the food truck scene. Chains from Chick-fil-A to Jack in the Box tried their luck on the road and launched their own food trucks." (Spiegel, 2014). On May 26, 2014 Starbuck introduced a "Frappuccino" truck in the Los Angeles and Long beach areas. This was followed by "TGI Fridays," the "Maille Mustard Mobile," the "Taco Bell Food Truck," and the White Castle "Crave Mobile."

Conclusion

Food in general has been seen as offering occasions to express identity and culture and in this way serve as a medium of communication. Mobile food vending has become a feature of daily urban life serving more than calories and sustenance. experiences for modern identity, culture, and politics. Food trucks and carts can be a feature of everyday life or a destination. They offer opportunities for exposure to symbols of diverse ethnic identities sharing the landscape of a city. The regulatory environment of mobile food vending addresses public health, public safety and public spaces. The primary aim of these regulations are aimed at non-communicative activity yet the communication functions of mobile food is affected. The National League of Cities' report recommended cities may need to reexamine their regulatory codes governing mobile food entrepreneurs (Food on Wheels, 2013).

The unintended consequences of such regulations should be added to calculations made in weighing the regulation of mobile morsels.

References

Bedeau, J. (2015). *Street food et food trucks a paris*. Hachette Livre: Vanves Cedex.

Berton, E. (2015, August 31). Food trucks finally arrive in food-fussy Paris, and people like them. *USA Today*. Retrieved from http://www.usatoday.com/story/news/world/2015/08/28/france-paris-food-trucks/31972603/

Business Improvement Districts, (2017). NYC Small Business Services. Retreived from http://www1.nyc.gov/site/sbs/neighborhoods/bids.page

Butler, S. (2014, August 8). From chuck wagons to pushcarts: The history of the food truck, history. Retrieved from http://www.history.com/news/hungry-history/from-chuck-wagons-to-pushcarts-the-history-of-the-food-truck

Greene, C. P. & Cramer, J. M. (2011) Beyond mere sustenance: Communication as food, In: J. M. Cramer, & C. P. Greene, C (Eds.). (2011). *Food as communication—communication as food pp. ix-xix.* *f*New York, NY: Peter Lang.

Davidson, A. (2013, May 7). The food-truck business stinks. *The New York Times Magazine.* Retrieved from http://nyti.ms/YAhbdV

Fabian, J. (2011, January 21). On Twitter, Obama touts D.C. food truck. The Hill.

Fickenscher, L. (2014, May 11). Food trucks stuck in park. *Crain's New York Business.* Retrieved from http://www.crainsnewyork.com/article/20140511/HOSPITALITY_TOURISM/140509816/food-trucks-stuck-in-park)

Field, H. (2016, January 25). 10 Absolutely amazing food trucks around the world cape town, Montreal, Paris, and Berlin are just a few of the cities on the food truck train. *The Daily Meal.* Retrieved from http://www.thedailymeal.com/travel/10-absolutely-amazing-food-trucks-around-world

Fisher, M. F. K. (1988). *How to cook a wolf.* North Point Press.

Food on Wheels: Mobile vending goes mainstream. (2013). National League of Cities publication. Retrieved from www.nlc.org/foodtrucks

Frommer, R., & Gall, B. (2012, November). Food truck freedom: How to build better food-truck laws in your city. Institute for Justice. Retrieved from http://ij.org/wp-content/uploads/2015/03/foodtruckfreedom.pdf

Gold, J. (2012). How America became a food truck nation. *Smithsonian.com.* Retrieved from http://www.smithsonianmag.com/travel/how-america-became-a-food-truck-nation-99979799/

Hoffman, B. (2011). A brief history of food vending in New York City: Midtown lunch. Retrieved from http://midtownlunch.com/2011/11/30/a-slice-of-street-vendor-history/

Kayden, J. S. (2000). *Privately owned public space: The New York City experience.* New York, NY: Wiley.

Lum, C. M. K., & Le Vayer, M. (Eds.). (2016). *Urban foodways and communication ethnographic studies in intangible cultural food heritages around the world.* Lanham, MD: Rowman & Littlefield Publishers.

Mobile Cuisine. Retrieved from http://mobile-cuisine.com/business/history-of-american-food-trucks/

Monroy v. City of New York. (2012, May 8). Retrieved from http://caselaw.findlaw.com/ny-supreme-court-appellate-divi-sion/1600535.html

Moskin, J. (2012, June 3). Food trucks in Paris? U.S. cuisine finds open minds, and mouths. *New York Times.* Retrieved from http://www.nytimes.com/2012/06/04/world/europe/food-trucks-add-american-flavor-to-paris.html

New York City Council Hearing on street vending modernization. (2016, October 27). CUNY Urban Food Policy Network. Retrieved from http://www.cunyurbanfoodpolicy.

org/news/2016/10/27/new-york-city-council-hearing-on-street-vending-modernization?rq=street%20vendors

New York City's street vendors work to decriminalize sidewalk selling. (2015, July 28). Free Speech Radio News. Retrieved from https://fsrn.org/2015/07/new-york-citys-street-vendors-work-to-decriminalize-sidewalk-selling/

Ngo, J. (2012). Mobile food vending and the public realm: A Seattle study (*Thesis University of Washington*). Retrieved from https://digital.lib.washington.edu/researchworks/bitstream/handle/1773/20787/Ngo_washington_0250O_10585.pdf;sequence=1

Norman, E., Frommer, R., Gall, B., & Knepper, L. (2011). *Streets of dreams*. Arlington, VA: Institute of Justice.

People v. Ala Carte Catering Co., 159 Cal. Rptr. 479 (Cal. App. Dep't Super. Ct. 1979).

Pittsburgh City Code §719.05A(d).

Rodriguez, C. (2015, April 24). Paris greenlights food trucks to feed city's fast-food craze. *Forbes*. Retrieved from http://www.forbes.com/sites/ceciliarodriguez/2015/04/24/food-trucks-drive-into-the-heart-of-paris-and-parisians-hearts/#520ee6c13043.

Rules and regulations for mobile food vending. (2016). NYC Department of Health and Mental Hygiene.

Shuttlesworth v Birmingham, 394 US 147, 150–151 [1969]).

SoHo Vending Guidebook. (2016). Retrieved from http://www.sohobroadwaybid.org/wp-content/uploads/FINAL-sidewalk_vending_guidebook-web1.pdf

Spiegel, A. (2014, June 11). 5 Companies that may drive food trucks into the ground. *Huffington Post*. Retrieved from http://www.huffingtonpost.com/2014/06/11/food-trucks_n_5481483.html. What mobile food vendors should know, NYC.gov

Swan, C. (2013, Oct. 19). Pushing privatized "partnership" agenda at New York City's public parks—Part 3. Retreived from http://www.huffingtonpost.com/cathryn-swan/pushing-privatized-partnership-agenda_b_3769191.html

Woodward, D. (2012, May 9). Rules of conduct. The architectural league's urban omnibus. Retrieved from http://urbanomnibus.net/2012/05/rules-of-conduct/

Zagat Interactive Food Truck Finder. Retrieved from http://mashable.com/2010/11/09/zagat-food-trucks/#VSTw7pWj2ZqR

Contributor Bios

David S. Allen is a professor in the Department of Journalism, Advertising, and Media Studies at the University of Wisconsin-Milwaukee. He is the author of *Democracy, Inc.: The Press and Law in the Corporate Rationalization of the Public Sphere* (University of Illinois Press) and co-editor of *Freeing the First Amendment: Critical Perspectives on Freedom of Expression* (New York University Press). His research on free speech law and ethics has been published in a variety of academic journals. His latest research focuses on the history and regulation of the public fora in democratic societies.

Emily Bauman is an Art History Ph.D. Candidate at the Institute of Fine Arts at New York University, concentrating in Modern & Contemporary Art and Architecture. She was educated in art history, theory and criticism at the School of the Art Institute of Chicago and Cornell University. Her dissertation is on the early performance and video work of Bruce Nauman, Paul McCarthy and Mike Kelley.

Gene Burd is a journalist, writer, and poet. He has more than 60 years as a newspaper reporter (Kansas City Star, Houston Chronicle, Albuquerque Journal, and suburban weeklies in Chicago, Michigan and Southern California); magazine freelancer and staff of Encyclopedia Britannica; government and media service in Chicago, Milwaukee and Denver; 50 years university teaching journalism and minority media at Marquette, Minnesota and Texas. Educated at UCLA (MA) and Northwestern (Ph.D.), and published in numerous interdisciplinary books and journals. He has provided endowments to establish the annual Media and Politics Institute in Los Angeles for Junior Statesmen of America and The Urban Communication Foundation in New York. He is listed in Who's Who in America.

Kevin M. Carragee is a Professor at Suffolk University in Boston. His scholarship focuses on contested urban space, the interaction between social movements and the news media, news and ideology, and forms of communication activism. He has edited six books and has published more than 25 journal articles and book chapters. His scholarship has appeared in leading journals in communication and journalism, including the *International Journal of Communication*, *Critical Studies in Media Communication*, *Journalism and Communication Monographs*, *Political Communication*, and the *Journal of Communication.* He has received a Fulbright Scholarship for teaching and research in Poland. He is a co-director of the Media Research and Action Project, a group that assists social movement organizations in framing their messages, influencing news coverage, and securing political change.

Charles M. Davidson is a Director of the Advanced Communications Law & Policy Institute (ACLP) at New York Law School. His academic focus is on the legal, policy, and regulatory reforms needed to spur broadband adoption and utilization across key demographics and sectors of the economy, to strengthen America's energy sector, and to modernize the nation's water infrastructure. Davidson holds a Masters of Law in Trade Regulation from New York University, a Masters in International Business from Columbia University, and B.A. and J.D. degrees from the University of Florida, where he graduated Phi Beta Kappa and served as a fellowship instructor at the College of Law.

Juliet Dee is an associate professor in the Department of Communication at the University of Delaware. She teaches courses in First Amendment law, mass media and culture, broadcast programming and television production. She has been director of the Legal Studies Program at the University of Delaware, has been an editor of the *Free Speech Yearbook,* and is a co-author of *Mass Communication Law in a Nutshell* (2014). She received her bachelor's degree from Princeton University, her master's degree from Northwestern University, and her doctorate from Temple University. She has published articles or book chapters on First Amendment issues involving controversial artwork funded by the National Endowment for the Arts, copyright infringement, anonymous defamation on the Internet, objectionable lyrics in rock music and hip-hop, media liability for violent content, media liability for classified ads for hitmen, and the Occupy Wall Street movement, hate speech and intentional infliction of emotional distress, and the issue of cyber-harassment and cyberbullying.

Susan Drucker, (Juris Doctor, St. John's University) is a Professor in the Department of Journalism/Media Studies, School of Communication, Hofstra University. She is an attorney, and treasurer of the Urban Communication Foundation. She has served as editor of the *Free Speech Yearbook*, *Qualitative Research Reports in Communication (Taylor & Francis)* and served as Series editor of the Communication and Law series for Hampton Press and Peter Lang Publishing. She is the author and editor of 13 books and over 100 articles and book chapters including two volumes of the *Urban Communication Reader*, *Regulating Convergence* (Lang, 2010), Voices in the *Street: Gender, Media and Public Space and two editions of Real Law @ Virtual Space: The Regulation of Cyberspace (1999, 2005)*, *Regulating Social Media: Legal and Ethical Consideration* (2013) with Gary Gumpert. She is currently completing *Urban Communication Regulations: Communication Freedoms and Limits*. Her work examines the relationship between media technology and human factors, particularly as viewed from a legal perspective.

Donald Fishman (Ph.D., Northwestern University) teaches courses in communication law and crisis communication at Boston College. He is the recipient of the Haiman Award for Distinguished Scholarship in Freedom of Expression and the Phifer Award for Outstanding Scholarship in Parliamentary Procedure. Fishman is the past president of the Eastern Communication Association, and he served as chair of the Department of Communication at Boston College for nine years.

Gary Gumpert (Ph.D, Wayne State University) is Emeritus Professor of Communication at Queens College of the City University of New York and President of the Urban Communication Foundation. His creative career as a television director and academic career as a scholar spans over 60 years. In 1960 he directed *the Gutenberg Galaxy* in which Marshall McLuhan articulated the premise of his forthcoming book. He is series editor of Urban Communication Series for Peter Lang Publishing. He has authored and edited books include *Talking Tombstones and Other Tales of the Media Age* (Oxford University Press), *The Urban Communication Reader* (Hampton Press), *Regulating Convergence* and *Regulating Social Media: Legal and Ethical Considerations* (Peter Lang). He is a recipient of the Franklyn S. Haiman Award for distinguished scholarship in freedom of expression (NCA), the Louis Forsdale Award for Outstanding Educator in the Field of Media Ecology (MEA), and in 2011 received The Neil Postman Award for Career Achievement in Public Intellectual Activity. His primary research and theory agenda focuses on the impact of communication technology upon social and urban space.

Harvey Jassem, Ph.D. (University of Wisconsin-Madison) is Professor Emeritus of Communication at the University of Hartford specializing in mediated communication, urban communication, communication law & policy, emerging media, and alternative media. He was Founding Director of the University of Hartford's School of Communication and Loyola University-Chicago's School of Communication, Technology and Public Service. He is a Founder and Director of the Urban Communication Foundation and has been a Scholar in Residence at New York University, MIT's Media Lab, a Fellow of the American Council on Education, and hosts a weekly jazz program on WWUH-FM.

Emily Long Vito is Assistant Director of Development at the Film Society at Lincoln Center. She was the founding Director of Communications and Development for The LAMP and led efforts raising over $3 million for community-focused media education programs and tools. Under her tenure, the organization has grown from serving a handful of students to over 900 annually, and she has helped to build out The LAMP's workforce development and professional development programs. Key partnerships cultivated include the Pinkerton Foundation, the Aspen Institute, Con Edison, Google, Sony and more. Emily also led the fundraising, development and marketing for the MediaBreaker and MediaBreaker/Studios platform, which is used by teachers in 47 states and 25 countries. She holds an M.F.A in Dramaturgy and Script Development from Columbia University, and grew up in the Midwest.

Faith Rose is a partner at O'Neill Rose Architects. From 2004 to 2014 she directed New York City's Design Excellence Program, overseeing projects ranging from museums and libraries to fire stations and garages.

In 2014 Faith was appointed Executive Director of New York City's Public Design Commission, which reviews capital projects on city owned property. While there, Faith built on the Commission's history of prioritizing the quality and excellence of the public realm, and worked to foster accessibility, diversity and inclusion in the City's public buildings and spaces. Faith has written and spoken widely about the design of public spaces.

Michael J. Santorelli is a Director of the Advanced Communications Law & Policy Institute (ACLP) at New York Law School. In that capacity, he oversees all aspects of the ACLP's research and scholarly writing. Immediately prior to joining the ACLP, Michael was the lead staffer for the New York City Council's Committee on Technology. Michael received his undergraduate degree, with honors, from Tufts University, and his law degree, with honors, from New York Law School.

Subject Index

A

B

C

D

E

F

G

H

I

K

L

M

N

O

P

Q

R

S

T

U

V

W

Y

Z

Names Index

A

B

C

D

E

F

G

H

J

K

L

M

N

O

R

S

T

W

Z

URBAN communication

Gary Gumpert, *General Editor*

Cities are inherently places of communication, meeting spaces for interaction and/or observation. The nature of any communication venue is altered by social and technological circumstances and the urban environment is altered, in turn, by changes in communication patterns. We need to understand relationships among these significant forces—communication, technology, and the urban, suburban, rural environment—as they shape each other. Communication systems and urban social systems can be examined at multiple levels as scholars and planners examine interaction in public spaces, neighborhood communication patterns, and urban systems of transport.

The focus of this series is on social relationships in a swiftly changing communication environment. Media coverage of urban issues, conflict resolution and contested urban space, visual communication, rhetorical dimensions of urban life, film and the city, journalism, the ethnic press, local media and public policy are just some areas of relevance. Volumes in this series provide a forum to explore and discuss the challenges created by the intersection of communication and urban life, focusing on what communication scholarship has to offer for enhanced understanding of cities and for the development of a public policy that takes into account communication needs and practices.

For additional information about this series or the submission of manuscripts, please contact the series editor, Gary Gumpert, at listra@optonline.net.

www.ingramcontent.com/pod-product-compliance
Ingram Content Group UK Ltd.
Pitfield, Milton Keynes, MK11 3LW, UK
UKHW020200290926
15471UKWH00005B/984

9 781433 146312